William Richard Gowers

Syphilis and the Nervous System

Being a revised Reprint of the Lettsomian Lectures for 1890

William Richard Gowers

Syphilis and the Nervous System
Being a revised Reprint of the Lettsomian Lectures for 1890

ISBN/EAN: 9783337250478

Printed in Europe, USA, Canada, Australia, Japan

Cover: Foto ©berggeist007 / pixelio.de

More available books at **www.hansebooks.com**

SYPHILIS

AND THE

NERVOUS SYSTEM

BEING A

REVISED REPRINT OF THE LETTSOMIAN LECTURES FOR 1890
DELIVERED BEFORE THE MEDICAL
SOCIETY OF LONDON.

BY

W. R. GOWERS, M.D., F.R.C.P., F.R.S.,

CONSULTING PHYSICIAN TO UNIVERSITY COLLEGE HOSPITAL, PHYSICIAN TO THE
NATIONAL HOSPITAL FOR THE PARALYZED AND EPILEPTIC, ETC.

PHILADELPHIA:
P. BLAKISTON, SON & CO.,
1012 WALNUT STREET.
1892.

PRESS OF WM. F FELL & CO.,
1220-24 SANSOM STREET,
PHILADELPHIA, PA.,
U. S. A.

PREFACE.

These lectures, delivered three years ago, are now reprinted on account of the frequency with which I find it necessary to refer to statements made in them, and the inconvenience of being obliged to refer a reader to the Medical Journals in which the lectures originally appeared. Two translated reprints have been published, and this renders their reproduction in the English language the more desirable. Moreover, I have taken the opportunity of carefully revising them, and have made a large number of additions. These, although not obtrusive or extensive, will, I hope, be found to increase the practical value of what is said, and may serve to bring the lectures up to the level of our present knowledge if they are below this in their original form. At the same time, their scope and character make the need for such additions insignificant. Their chief object is to enable those who read them to grasp more firmly the cases they meet with, and to understand better the methods of dealing with the disease in practical thought and actual work.

My thanks are due Dr. D. D. Stewart, of Philadelphia, for assistance in reading proof, and for preparing the index.

Queen Anne St., London,
October, 1892.

TABLE OF CONTENTS.

LECTURE I.
THE ULTIMATE PATHOLOGY OF SYPHILIS.

The Actual Pathological Position and Macroscopic (non-histologic) Character of Syphilitic Tissue-formations ; Their Division into Classes,—*Specific* and *Non-special*.—General Consideration of These, Their Situation and Effects, Direct and Remote ; the *Specific* or Neoplastic, Consisting of Gummata and Growths in the Arterial Walls, and the Pathologically *Non-special* or Inflammatory Lesions.—The Question of Evidence as an Aid to the Diagnosis of Inflammatory or Pathologically *Non-specific* Lesions.—Evidence of Sequence and that of Therapeutics as an Aid to the Proof of Causation.—The Relations of Nerve-degenerations to Syphilis.—Analogy Between the Effect of Certain Poisons, Chemical and Organized, and that of Syphilis.—Strümpell's Hypothesis, ...   9–54

LECTURE II.
THE ORIGIN OF FUNCTIONAL NERVOUS DISORDERS ATTRIBUTED TO SYPHILIS ON IMPERFECT EVIDENCE.

Considerations in Outline of the Possible and Probable Diagnosis of Syphilis of the Nervous System Determined by the Situation and Nature of the Lesions, and the Character and Course of the Symptoms.—Symptomatology and Diagnosis of Some Syphilitic Processes Affecting the Nervous System, such as Gummata, Chronic Local Cerebral Meningitis, and Arterial Disease with its Sequences, Thrombosis with Necrotic Softening of the Brain.—Ocular Palsies.—The Importance of Accuracy of Diagnosis in Syphilitic Disease of the Nervous System, ...   55–90

LECTURE III.

THE ESSENTIAL PRINCIPLES UNDERLYING THE PROGNOSIS OF SYPHILITIC DISEASE OF THE NERVOUS SYSTEM, AND THEIR EFFECT UPON THE SPECIAL PROGNOSIS OF THE CHIEF LESIONS.

The Prognosis a Matter of Special Consideration of Individual Cases, as is also Diagnosis.—Symptoms of Syphilitic Disease of the Nervous System Depend not on the Specific Process, but on the Simple Alterations this Produces in Nerve-tissue. —As Treatment Exerts a Direct Effect on the Specific Process Only, the Persistence of Symptoms Despite Treatment Indicates the Persistence of Damage to the Nerve Elements, and not of the Luetic Lesion Producing the Damage.—Special Prognosis of the Chief Luetic Affections.—Modifications of Prognosis Entailed by Symptoms of Irritation Caused by Certain Lesions.—Syphilis an Incurable Disease—this Statement Consistent with the Recognition of the Fact that the Lesions of Syphilis are Readily Removed by Treatment.— Respective Power of Mercury and of Potassium Iodide in the Treatment of Syphilitic Diseases of the Nervous System.— How these Drugs are best Employed.—Efficient Measures of Prevention and of Arrest of Syphilis not yet at Hand.—Unbroken Chastity the only Certain Prevention, ... 91–127

INDEX, ... 128

SYPHILIS

AND THE

NERVOUS SYSTEM.

LECTURE I.

Mr. President and Gentlemen,—Permit me first to tender to you, Sir, and the Council, my thanks for selection to the post of Lettsomian Lecturer to this Society, a post around which cluster so many honorable associations. The duty which that selection lays upon me, and which I have now to attempt to discharge, has difficulties of its own. This Society constitutes an audience with the widest professional interests, but above all with an interest in the practical aspects of medical science. This characteristic has been handed down through successive generations of its Fellows, from the time when theory bore a much smaller proportion to practice than it does now. It is not one that should be lightly esteemed or hastily discarded, especially in these days, when the "scientific imagination" is apt to wander far and free. The subject to which I propose to ask your attention is one in which, more perhaps than in any other, the two great branches of the profession join hands in mutual supplement of work. This co-operation is the more important because the disease we are to consider turns to each a different face. To the surgeon the processes of syphilis are for the most part open and manifest; to the physician they are secret; its ways are obscure, its language is seldom unequivocal.

This subject has, indeed, once already been made that of the Lettsomian Lectures. Those of my distinguished predecessor, Dr. Broadbent, together with the almost simultaneous work of Dr. Buzzard, were the means of first diffusing widely, in this country, a knowledge of the extensive influence of syphilis on the nervous system. But the fifteen years that have since passed have been prolific in additions to our knowledge, and the time seems ripe for a fresh review of at least some aspects of the subject. Not only are there many new facts that deserve notice, but there are many doubtful points that need consideration. If it is desirable to review our positive knowledge, it is not less important clearly to define or even to discern its limits. There is perhaps no subject on which there is so much of what may be termed (if the convenient contradiction is permitted) uncertain knowledge—in which there are so many opinions that rest on no secure foundation, and yet are accepted as true—so many mere hypotheses that are put forward and received as more than hypothetical; so much that is possibly true, and yet is still unsupported by any real evidence. Hence it seems especially desirable to consider the evidence that has been or may be obtained as to the relation to syphilis of many of the diseases of the nervous system. In doing so, we shall learn something of them, and see how we may learn more. There are also practical questions in diagnosis and prognosis that need consideration. Lastly, there are questions in treatment on which current opinion is divided, in some cases widely divided; a discussion of these, even if partial, may be of service by drawing to them the attention of the many, which they at present receive only from a few. These, then, are the topics to which I propose to ask your attention.

I dare not depart from the time-honored custom of a preliminary apology, especially as I feel that an apology is not unneeded. Indeed, that I may be quite on the safe side alike of custom and of conscience, I will be bold in my humility, and offer you two apologies. First, you have a right to expect from a Lettsomian lecturer something new, some definite addition to our knowledge. This reasonable expectation I fear I must disappoint. It is scarcely compatible with the chief object that I desire to attempt. My second apology arises from the fact that we cannot separate one part of a disease from the rest. In considering syphilis as it affects the nervous system, and especially the question of evidence, I cannot avoid alluding to points in its history that I am not competent to discuss. In this also I must ask your indulgence, and it may save time and words if, once for all, I beg you to impute to me all due diffidence and hesitation wherever they are seemly, even though, in the exigence of the moment, they may not be manifest.

Among the visceral effects of syphilis those on the nervous system stand first in extent, variety, frequency, and gravity. Considering this fact, it may seem strange that our knowledge of these lesions should have been of such slow growth and late development, and that so vast a range of syphilitic influence was scarcely suspected thirty years ago. But we must not forget that thirty years hence some future Lettsomian lecturer, now playing with his toys, may make the same remark of the knowledge of the present day. The cause that has hindered our acquaintance with these diseases is not far to seek. It is a cause that we cannot and must not hope to be lessened in the future—the old and inevitable antagonism between pathology and therapeutics. Each flourishes at the expense of the other. The ends for which we gain our knowledge are the chief obstacle to its acquisition. We know most about the pathology of the diseases that we fail to cure. In proportion as we are successful in treatment we are dependent on the "chapter of accidents" for information as to the real nature of the disease. This influence has affected syphilis more than most other maladies.

The "ultimate pathology" of syphilis, even in its special relations, is beyond the limit I have set to the subject of these lectures. That the cause of syphilis is a micro-organism can scarcely be doubted by any one who has followed, even at a distance, those discoveries that have transformed so much of our pathology, although it may reasonably be doubted whether the organism itself has yet been detected. It is also highly probable that the organism is the immediate cause of most lesions, which can be promptly influenced by the treatment known as "anti-syphilitic." But beyond this I must not attempt to discuss the pathology of the disease. It was part of my original intention to endeavor to confer some semblance of vitality on the dry bones I have to offer you, by indicating the form in which some of the pathological problems of syphilis present themselves at the present day, and the kind of answer we may reasonably expect from the knowledge that we hope the near future has in store. But for a profitable discussion of such questions, however brief, the time is not yet ripe. At present we have no guide but analogy, certainly not close, and possibly misleading, and I desire to avoid the region of mere analogy and of pure speculation.

Permit me, however, at the outset, to remind you of certain general considerations regarding the effects of syphilis,—considerations that are not speculative, but are merely abstract and familiar to you, certainly in fact, and probably in form. If we compare the various syphilitic lesions, not limiting our view to any single organ, we may discern in them two elements—a process of inflammation, and a process of tissue-formation. It might seem more accurate to call the former the "congestive" element, since tissue-formation is so often a part of the process of inflammation, but the term "congestion" involves a limitation, and is scarcely wide enough for the purpose. We can mentally distinguish these two elements, and the distinction is aided by the fact that they are combined in various degrees in the different lesions. The process of tissue-formation is seen, in almost a pure form, in syphilomata,—"gummata," as they are called from their gluey aspect. In these no more of the congestive element is to be discerned than in most other tumors. On the other hand, the inflammatory or congestive element is almost or quite pure in many of the early skin eruptions, in some cases of iritis, and in many other lesions. Generally the two elements are combined, and the most common condition is that often termed "hyperplastic inflammation." Sometimes the combination is such that it is not easy to decide in which class a lesion should be placed, whether it should be regarded as a hyperplastic inflammation or as a diffuse gumma.

If we survey the course of active syphilis, there seems to be a tendency for the congestive or inflammatory element to preponderate in the early lesions, and the tissue-formation in the later. At the same time, the earliest lesions, the primary induration and the enlargement of the nearer glands, are apparent exceptions to this law—exceptions that will have to be considered whenever wider and deeper knowledge brings with it the possibility of framing an adequate hypothesis as to the real pathology of the disease, and the relation of the virus to the several lesions.

It has, indeed, been suggested that there is no real difference between these two elements. The tissue formed in syphilitic processes differs but little from that formed in ordinary inflammation, and inasmuch as it is combined in varying quantity with the signs of inflammation, it is urged that even the gumma has closer affinities with inflammatory products than with the true new growths. But the admission of this opinion (if it is justified) does not preclude the distinction of the two conditions, for it is often well to distinguish when we have no right to divide. The distinction seems to be justified if each condition may exist so far alone that we cannot clearly discern the other.

Whatever be the real pathological position of the tissue formed, it presents certain characters not found in the same degree in ordinary inflammatory products, and it is very important to keep these in mind. First, the peculiarity of aspect when of recent formation, to which the name "gumma" is due. Secondly, the tendency to undergo certain changes—for the tissue-elements to perish in caseation, or else to undergo a change into a dense contracting fibrous tissue. The gluey aspect and the caseation constitute the most important characters by which this tissue can be identified. The tendency to degeneration is so widely but unequally diffused through a mass, that the cheesy change usually begins at many places. This tendency to multiple caseation, and the irregular areas thus produced, constitute the most important criterion of the nature of the tissue, when the "gummy" aspect has ceased to be distinctive. Lastly, the tendency to change to a contracting fibrous tissue is of extreme importance in relation to the effects produced—by which the lesion is manifested. This fibrous change often takes place in the tissue at the surface of a gumma, as part of its degenerative or cicatricial changes, and it is probably this that has given rise to the erroneous statement sometimes made, that a capsule is characteristic of these growths. It may also occur throughout the new formation, as in the inflammation of the dura mater which surrounds the spinal cord with a sheath of this dense tissue.

We may divide syphilitic lesions into certain classes, according to our knowledge of their pathological characters, and we may expect to find these classes represented in the nervous system. First, there are morbid processes with characters that are peculiar, processes due to no other cause. Secondly, there are processes that are not peculiar; so far as we can see, they resemble those that are due to other causes, although they are produced by syphilis. Some of this second class, however, although not special in pathological characters, are yet, when due to syphilis, peculiar in their seat and distribution. Others have not even this peculiarity, but may be described as lesions due to many causes, one of which is syphilis. I do not say they are not really special and peculiar; I only say that we cannot discern the peculiarity. Outside the nervous system we see many examples of these varieties. They are conspicuous, for instance, in the skin. A gumma, ulcerating in a serpiginous manner, is an example of a truly special lesion. Many other cutaneous syphilides, such as roseola and psoriasis, however, have no pathological peculiarity, although most of them, when syphilitic, are peculiar in distribution. Many cases of syphilitic iritis are special in aspect, but others present no peculiarity to distinguish them from inflammation of the iris due to other causes. Thus simple iritis, simple so far as we can see, has several causes, of which syphilis is one. Speaking generally, the more the congestive or inflammatory element preponderates in a lesion, the less special are its pathological features. These depend chiefly on the element of tissue-formation. Necessarily also, as the latter needs time for its production, the more acute a syphilitic inflammation is, the less are its features distinctive. This is, I think, a very important consideration.

It is desirable to keep this classification in mind in considering the lesions of the nervous system, because the evidence as to the syphilitic nature of a lesion necessarily differs in the various classes, and it will be desirable to adopt this classification in reviewing the several lesions and the evidence regarding their nature. I must pass unnoticed those that are caused by bone-disease.

We begin, then, with the lesions of which syphilis is the only cause. They possess distinctive characters, and it is found that syphilis is their invariable antecedent. It can be traced in a large proportion of the cases, and the more thorough our knowledge of the individuals, the smaller is the proportion in which the evidence of syphilis is not clear. On the other hand, those lesions are never met in cases in which syphilis can be excluded with certainty. From their special characters they may be termed "specific" lesions. We need this term for them, and it is a misfortune that the word has come to be used in a wider sense, as a euphemism for "syphilitic." For this, moreover, there is a word which might with advantage become as current here as it is in Germany—"luetic." I propose, therefore, to use the term "specific" in the sense of a lesion that is special, not only in its causation, but also in its characters.

These truly "specific" lesions I must pass over with merely a brief mention, because they are the best known of the effects of syphilis, regarding which there is least room for question. It will be remembered that I am now dealing only with pathology; we shall have again to consider these lesions in regard to practical diagnosis, in which they furnish many important problems.

First, there are the syphilitic growths, gummata, circumscribed masses of new tissue having the characteristics I just mentioned. They grow commonly from the pia mater, and both compress and invade the substance of the brain and spinal cord. They are rare within the substance of the brain, and they are rare also, although sometimes symmetrical, on the cranial nerves. Occasionally there is a more diffuse formation of similar tissue in connection with the dura mater, cerebral or spinal.

Secondly, there are the growths in the walls of the arteries, which constitute one of the most important of these specific lesions; but they also are well known and need not detain us. One point, however, should be mentioned. As generally met with the disease is recent and consists in a thickening of the wall which is more limited than in atheroma and less opaque. But when such a disease has been treated, as by a course of iodide of potassium, the thickening is less and the opacity greater, so that the condition resembles more closely ordinary atheroma, for which it may usually be mistaken. I have seen such disease in one artery associated with recent and characteristic changes in another. I would ask you to note, also, in passing, the remarkable correspondence in seat between this arterial disease and ordinary atheroma. Except in the arteries of the brain and in the aorta, syphilitic disease is scarcely known. We ascribe the proclivity of those arteries to atheroma to the influence of strain; is this also the cause that determines their affection in syphilis? I can think of no other, but the fact must be borne in mind in connection with certain theories of the origin of the lesions.

Many facts regarding this arterial disease will have to be considered when we come to the diagnostic problems they furnish, but certain of its pathological consequences must not be passed over. The first is the necrotic softening that follows occlusion of the diseased artery, or more often that of a branch coming off from the diseased spot. It is by this necrotic softening that the disease most frequently causes symptoms. Occlusion, however, does not always lead to softening. Induration may result, especially in the cortex, and more often in children than in adults. It is no doubt due to the arterial anastomoses being sufficient to prevent actual necrosis, although insufficient to maintain the integrity of the tissue.

A second effect is the production of aneurism of the

larger arteries, those that are the special seats of the disease. Such aneurisms seem to be always due to inflammation of the wall, by which extensible fibroid tissue is substituted for the normal elastic elements. We know two other causes of such intracranial aneurisms. One, very rare, is traumatic arteritis. The other is embolism, imperfectly occluding the vessel; this is probably the cause of two-thirds of such aneurisms before the degenerative period of life. In the remainder, in which there is no history of injury and no evidence of embolism, there is a history of syphilis in so many cases as to justify the opinion that most of them are due to this influence. Often the history is imperfect, because the aneurism has been unsuspected until the final rupture and syphilis has not been enquired for. When we consider how great is the amount of new growth in the wall of a diseased artery, and how prone are the new elements to change into extensible fibroid tissue, the wonder is that aneurism is not a more frequent occurrence. Probably the explanation is to be found in the common persistence of the elastic layers, which afford the chief safeguard against permanent dilatation. I am only able to offer one additional instance of an aneurism due to syphilis. A girl, aged about twenty-five, who had suffered from syphilis, was admitted to University College Hospital with the symptoms of a small tumor near the crus. Soon after admission, she suddenly became unconscious, and died in a few hours. A large amount of blood was found at the base of the brain; it had come from a small aneurism, the size of a large pea, on the posterior cerebral artery, near its origin. No source or evidence of embolism could be found.

A third, but very rare, effect of syphilitic vascular disease is hemorrhage within the substance of the brain. Doubtless, the reason why it is so seldom met with is because the arteries within the substance of the brain are so seldom the seat of syphilitic disease. A few cases have been met with, and I have seen one remarkable instance of it in the inherited malady. The patient was a boy, aged twelve, with clear evidence of congenital syphilis. He had suffered from some transient cerebral symptoms at the age of eight. The hemorrhage was a large one; it had destroyed the greater part of one cerebral hemisphere and had burst into the ventricles. Although its source was not discovered, there was such disease as to leave no doubt of its cause. Extensive and characteristic syphilitic disease existed in the right vertebral artery, and some smaller spots were seen on the posterior cerebrals. Moreover, the aorta presented some patches of similar disease, which was conspicuous also on both coronary arteries, near their origin and along their course, while several raised yellowish spots existed on the flaps of the mitral valve. There were no vegetations, and it should be added that the boy had not had rheumatic fever. The case is thus remarkable for more than one reason, not the least being the late stage of the inherited disease at which the lesion occured.

Inflammation that is specific in its pathological features is chiefly meningeal. The special feature in this also is its hyperplastic character, much new tissue being produced, which may undergo the caseous or the fibroid degenerative change. Either of the two membranes may be inflamed. In the dura mater the inflammation is usually diffuse, and the tissue formed undergoes chiefly the fibroid change. Such cerebral pachymeningitis is more common in the inherited than in the acquired disease. In the pia mater the inflammation is usually local. The new tissue may undergo the characteristic caseation, or fibroid change in it may result in the formation of a layer of dense cicatricial tissue, in which nerve roots at the part are inseparably involved.

In the spinal cord the most frequent chronic inflammation is that of the dura mater, the important and well-known "spinal pachymeningitis." This may damage greatly not only the spinal and bulbar nerve roots, but the cord itself, from the great amount of fibrous tissue produced. The pia mater is seldom inflamed locally, and its diffuse inflammation is usually combined with that of the dura mater.

Inflammation of the substance of the brain or cord that can be regarded as pathologically specific is very rare. Primary focal softening and induration have been met with, but most of the cases are open to the question whether the lesion was not due to arterial disease. When these changes do result from inflammation, they are seldom specific in their characters. A case of chronic disseminated inflammation has been described by Charcot and Gombault, which may, however, be regarded as a typical example of such a specific inflammation. Foci of chronic inflammation existed in the brain, pons, and optic nerves, and in many of these foci the new tissue presented multiple points of caseation, although in others it had undergone only the fibroid change.

Very little is definitely known of inflammation of the cranial nerves which is pathologically specific. These nerves are often damaged by gummata and by chronic meningitis, but there is also reason to believe that they are sometimes the seat of hyperplastic inflammation, involving the sheath and interstitial tissue. The distribution is usually irregular, and it is seldom bilateral. It is remarkable that the peripheral parts of the spinal nerves are very seldom affected. The discovery of the susceptibility of these nerves to various toxic influences has revealed no liability for them to suffer in syphilis.

It will be observed that in speaking of specific pathological characters I have alluded only to those that can be recognized by the naked eye. I do not think that, at present, any confidence can be placed on histological characters only. Even the structure of the gumma is not distinctive; all that the microscope enables us to say is that the tumor *may* be syphilitic. In the case of inflammations, it is very doubtful whether the microscopical characters have anything like the significance sometimes ascribed to them; all the characters need to be considered.

Are there any lesions of the nervous system that can be said to belong to the second class—lesions that present no pathological peculiarity—lesions that are due to other causes besides syphilis, but which when syphilitic in origin have peculiarities of distribution not present when they are due to other causes? There are a few that may possibly be of this character. One is a form of disseminated subacute myelitis, in which defined islets of inflammation occupy the white substance of the cord, chiefly near the surface, and are perhaps sometimes met with in the brain. This form has been described, as related to syphilis, by Julliard and Pierret, and I have met with one instance. Although syphilis seems to have been its antecedent in the cases hitherto published, more observations are needed before it can be regarded as a specific lesion, even when its peculiar insular character is taken into consideration, and the facts at present do not prove specificity apart from its distribution. The same may be said of a form of multiple inflammation of the spinal roots, regarded as a specific by Kahler, and of certain forms of sclerosis of the cortex of the brain (without arterial disease) met with by Barlow in infantile syphilis.

It is not likely that this completes the series of obviously specific lesions. Many other morbid processes may yet be added, by reason of their characters or seat, when our knowledge of them is rendered adequate by a sufficient number of observations. For the present, however, they must be placed in the class we have now to consider, the lesions of which syphilis is one of several possible causes—lesions analogous to the apparently simple iritis. The chief morbid processes which seem to fall into this class are two: first, certain acute and subacute inflammations; and secondly, certain forms of chronic inflammation or focal sclerosis of the nerve-centres. In the same category, moreover, we must also place the important class of primary nerve degenerations. These we shall notice later, but the methods we are about to consider are applicable to them also.

The question of evidence is of the greater importance in regard to this class because we have to proceed to some extent without the aid of pathological observation. We can make the diagnosis during life with considerable confidence in some of these cases, especially in the acute inflammations. In them, opportunities for pathological observation come seldom, and it is very doubtful whether, when they do come, they afford any real help. We have already seen that it is the new tissue which gives to the *specific* lesions their distinctive peculiarity. For the formation of this tissue, time is essential. In proportion as syphilitic inflammations are acute, the inflammatory or congestive element preponderates, and the lesion assumes an apparently simple character. Obviously, therefore, if an acute inflammation is produced by syphilis, we cannot expect it to present such specific features as would declare the nature. We cannot expect, therefore, that a post-mortem examination would decide the question; and yet we need, and urgently need, to know whether such lesions are or are not among the results of syphilis. But in the case of these apparently simple lesions, only two kinds of evidence seem at present within our reach: the first is such sequence as proves causation; the second is the evidence of therapeutics. It is desirable to examine these two kinds of evidence, to inquire how they can be obtained, and to what limitations they are subject. This inquiry is important, because current ideas on the subject seem ill-defined, and yet we have no other evidence. I ought, indeed, to recognize a possible future source of definite knowledge most part vague (at least if we may judge by their application), and also because, with one exception, I do not see where else we are to look for evidence.

The progress of bacteriology may supply us with a conclusive criterion. Such a criterion would be available also for the specific lesions, but for them it would be merely confirmatory; for the lesions now under consideration it would be essential. Unfortunately, such a criterion is still in the dim, if not in the distant, future. We cannot assume that it will certainly follow the discovery of the specific organism, or that its application would be unlimited. It is not likely that the test would be of negative as well as positive significance,—that the absence of the organism would be as conclusive as its presence. Thus even this criterion may not relieve us from the need for other proof.

The evidence of sequence is thus of great importance. Regarding it, the first important consideration is that isolated instances are no proof of causation. (Let me again remind you that I am now speaking of scientific proof, not of practical diagnosis.) Individual sequence would be proof if we could exclude all other causes of a disease, but this is, as a rule, beyond our power. The diseases in question have many causes, and those causes are imperfectly known, so that their exclusion in a given case is impossible. The fact that we cannot trace them does not prove that they do not exist. We are often unable to trace the cause when there is no question of syphilis, although some adequate cause there must be. Hence, when there is syphilis, the fact that we cannot trace any other cause does not justify us in concluding that the disease in question is due to syphilis. To do so is to treat the unknown as non-existent, and to base our induction not on knowledge, but on ignorance. Consider how many diseases coincide with, or succeed, constitutional syphilis, which no one dreams of attributing to it. We do not ascribe acute rheumatism or acute sthenic pneumonia to preceding syphilis, and yet, because a morbid process occurs in an organ, as the spinal cord, which is sometimes the seat of syphilitic processes, it is assumed that the process is due to syphilis, merely because the patient is or has been syphilitic. It is perfectly certain that such a combination must often occur when there is no causation. It is certain, from the mere law of chance; but such a coincident, without *direct* causation, may well be more frequent than mere chance will explain. A general disease like syphilis may so influence the general health as to predispose to many diseases for which other causes may exist. Therefore, while individual instances of sequence may justly direct attentention to the possibility of causation—while they may suggest a search for other evidence—they do not prove causation. Indeed, I should not venture to call your attention thus specially to a rule that is so obvious, were it not that medical literature abounds with cases in which causation by syphilis is assumed to be proven by a single sequence.

How, then, can sequence prove causation? How can the fact that a disease follows syphilis be evidence that it is due to syphilis? Only by the strength of numbers. The sequence that is without significance when isolated, has the highest significance when traceable in a series of cases. This fact is so well known, and the method has been employed so frequently in the case of tabes and other maladies of the same class, that it may seem needless to consider it in detail. But its use must be widely extended if we are to ascertain whether or not many other diseases are related to syphilis, and its value depends largely on the manner in which it is used. The number of cases necessary will vary according to the field of observation, the frequency of the disease, its incidence on age and sex, and the nature of the proof required. A small number of coincidences, in a fair field of observation, might be sufficient to prove the mere fact of causation, although a much larger number would be necessary to enable its extent to be estimated. Used with care the method is of the utmost value; carelessly employed it is worthless and merits the sneers that have been so freely bestowed upon it.

It is only practicable to approach the question from the side of the disease to be tested, because even those diseases that are certainly consequences of syphilis are rare consequences, so that, from the side of syphilis, it is not easy to get an adequate view of them. It will also be convenient, in the case of many diseases, to restrict the inquiry to the age at which syphilis is common and the sex in which it is most readily ascertained. If the restriction is the same in each series compared, the validity of the result is not affected. It is, therefore, wise to let the cases compared be those of males only, and over twenty-five years of age. In some cases it may be well to narrow the limit to cases between twenty-five and fifty.

The first task is to ascertain what is the average proportion of males over twenty-five who have suffered from syphilis. This term of the comparison (which we may term "the standard") will be the same for all diseases. It is of great importance that it should be ascertained, once for all, from a field sufficiently extensive to place the conclusion beyond the reach of doubt. Although one standard will be sufficient for all diseases, it will probably not be sufficient for all classes or for all places, since it is probable that in both class and place the frequency of syphilis varies. It is probably greater in the lower than in the middle and upper classes, greater in urban than in rural districts, greater in seaports than in inland towns. Attempts have been made abroad on a small scale to ascertain the proportion, but not, I think, in this country. I regret that I am able to offer only a very small contribution to it. I venture, however, to hope that the need, once pointed out, may not be long unsupplied. Messrs. Raymond-Johnson and Marriott have kindly ascertained the venereal history, positive and negative, of 112 cases at University College Hospital, and Mr. Littlewood that of 200 patients at the Leeds Infirmary. The result is that in London there is a history of syphilis (including in a few cases a certainly infecting sore only) in 10.5 per cent., and of a chancre of uncertain nature in another 10 per cent. At Leeds the percentage of syphilis, or hard chancre, is only 6.4, and of a chancre of uncertain nature only 6 per cent. This shows very clearly the need for more than one standard, but the want might be supplied without much difficulty. A few months' observation by the registrars of all the London hospitals would supply us with a London standard, drawn from at least five thousand cases, which would be available for all similar inquiries for many years to come.

But the figures I have just mentioned place before us at once one of the difficulties of the process of comparison. It is certain that some of the patients, who give a history only of a chancre, have had an infecting sore—but how many? What proportion of the chancre-percentage should we add to the syphilis-percentage to obtain the nearest approximation to the truth? Mr. Berkeley Hill and others, in this country and abroad, have published analyses which show that at least two-thirds of all venereal sores are infecting. We have, however, already taken count of the cases of *certain* syphilis, and of the remaining sores probably the majority are not infecting. If we add to the percentage of syphilis one-quarter of the cases of chancre, it will give us about 13 per cent. of syphilis in London, and 8 per cent. at Leeds.

The proportion among the middle and upper classes

is probably less than among the poor, and I doubt whether it exceeds 10 per cent., even in London. The difficulty of ascertaining it will be much greater than among the poor.

In the other terms of comparison we ascertain in a similar manner the percentage of syphilis among the sufferers from the disease in question corresponding to the standard in sex and age. If the proportion exceeds that of the standard, the excess must be due either to the one disease being a cause of the other, or to the fact that both are due to a common cause. Obviously, however, the only conceivable relation is that syphilis is in some way the cause of the disease in question. If the excess is slight, and if the interval is short, indirect causation is possible; that is, syphilis may have predisposed merely as any other depressing disease may predispose. If the excess is considerable, we must assume direct causation.

But the comparison itself is not quite a simple matter. The significant fact is the *excess* of the syphilitics among those who suffer from the disease, and we have to ascertain this excess. It may seem sufficient to deduct the percentage in the standard from the percentage in the disease, and to consider the difference as the percentage of causation. But this would be erroneous, as an example will show. Let us assume, taking simple numbers for convenience, that the percentage of syphilis in the standard is 10, and in the disease 50. It is not right simply to deduct 10 from 50, and assume 40 per cent. causation. We start with 100 cases, and if we thus assume 40 to be the proportion of causation, we have left 10 cases of accidental coincidence and 50 without syphilis. Thus we assume that in 10 cases out of 60 there is mere coincidence, whereas our assumed standard shows that in 60 persons there will be only 6 with merely coincident syphilis. The true method, I take it, may be expressed in the following rule: Deduct from the syphilitics the number that bears the same proportion to the non-syphilitics as is found in the standard; the residue will represent the causation. Thus in our example we must deduct from the 50 syphilitics not 10, but 5.5, because not 10 but 5.5 bears the same proportion to 50 as 10 does to 90 (10 *in* 100), and the proportion in which there is causation will not be 40, but 44.5. It is certain that such a deduction for mere coincidence is necessary in the case of diseases that are often due to other causes, although it is open to consideration whether this deduction is justified in the case of diseases of which syphilis is the chief cause. But its amount lessens in inverse ratio to the proportion in which syphilis can be traced, and when this proportion is high, the deduction becomes so small as to be unimportant. If, for instance, syphilis can be traced in 80 per cent. of the cases, the deduction for coincidence would be less than 3 per cent. The error from neglecting it will then be scarcely appreciable, especially in the face of the graver source of uncertainty we have next to consider.

Whatever proportion of syphilis is found will fall short of the truth. This is certain from the fact that persons occasionally suffer from syphilis and do not know it, or know it only with a knowledge which soon passes from their memory. Many persons who present unequivocal *signs* of old syphilis can give a history only of a chancre, and others not even of this. We meet with persons who know of no sore, who give no history of syphilis, and yet have had the disease. This fact is well known, but it is so important that I venture to emphasize it with a few instances. A gentleman had once, and only once, a slight sore. He showed it to a doctor, who said it was of no importance, and touched it with caustic. In a few days it was well, and was soon almost forgotten, for no after-symptoms occurred to keep it in mind. In a year or so he married a lady in his own station: She bore him two dead children, and then a living one. The latter soon showed signs of syphilis, and became paralyzed and imbecile. Then the wife herself died from the effects of the same disease. Similar cases, we may hope less tragic, must be known to all of you—cases in which the transmission of the disease to others is the first proof of its existence. A striking instance is the first case of progressive ophthalmoplegia in which the nature of the lesion was proved. I suppose there is no one more skilled in tracing syphilis than Mr. Jonathan Hutchinson, and yet in this case the search, many times renewed, failed to reveal any indication of the disease; there was not even a venereal sore to suggest it; but after the man's death a child of his was brought to Mr. Hutchinson with interstitial keratitis and notched teeth. I have met with many patients who had signs of the past disease, in some indubitable lesion of eye, tongue, throat, or skin, who could give no history of a chancre. But I am able to offer you more definite facts, for which I am indebted to my friend Dr. Radcliffe Croker. He has been kind enough to ascertain from his case-records at University College Hospital the proportion of cases of tertiary syphilitic skin eruptions in which there was no history of a chancre. He has found 56 cases of such eruptions, cases that are consecutive and unselected, and in no less than 11 of these the occurrence of any chancre was unknown. This is a proportion of 20 per cent. It cannot be doubted, moreover, that besides such cases there must be many more who have no late symptoms whatever to reveal the disease.

These facts suggest a question most important in regard to the problem we are considering, the actual frequency of what may be termed *latent* syphilis, in which the manifestations of the disease are insignificant, and are either unnoticed or untreated, so that, years afterward, the memory of them is reduced to a vanishing-point in the vista of the past. May not such cases be far more frequent than is generally thought— more frequent, indeed, than we can ascertain? Besides the facts I have mentioned there are others which, if they do not suggest this conclusion, at least harmonize with it and are explained by it. Many observers have noticed how often severe late lesions have followed an early stage that was slight or unnoticed. It has, indeed, more than once been suggested that there is an actual alternation in severity between the early and the late effects. Such an alternation would explain the facts, but cases are frequent enough in which both stages are alike severe. May not the frequent apparent alternation be simply due to the great frequency of mild or latent syphilis? Take, again, the fact epitomized in what is termed (abroad more often than in England) Colles's Law —the fact that a woman who bears a syphilitic child, although apparently free, is proof against infection, and must therefore have suffered. It is supposed that pregnancy is in some way antagonistic to the activity of the disease, but such antagonism is certainly often not evident, and the anomaly may be equally explained by the frequency of latent syphilis.

It is therefore certain that the facts we can ascertain will fall short of the real facts, and will fall short of them by a quantity which, because it is unknown, must not therefore be neglected. Whatever the actual amount of the discrepancy, its influence will be the greater the larger the number of cases in which syphilis can be traced. For instance, to take what may or may not be an extreme proportion, let us suppose that syphilis cannot be ascertained in one quarter of the patients who have suffered from it—that, besides the cases in which there is clear evidence of the disease, there are onethird more, on an average, in whom we cannot ascertain it after the lapse of years. This would raise the proportion of syphilitics among adult males to about 18 per cent., but it would raise a proportion of 50 per cent. to 70, and of 60 per cent. to 80. To ascertain the mere fact of causation we may indeed neglect the discrepancy, but we must recognize, at least its existence, when we endeavor to estimate the extent of causation.

One last general remark. For the comparisons to be significant the facts must, as a rule, be specially observed. It is one of the many lessons of scientific history, in no part more conspicuous than in medicine, that facts observed without special reference to the question at issue are of very little value for its decision.

On the second kind of evidence, that of therapeutics —the influence of drugs on the disease—I must, and can, be more brief. The considerations are not less important, but they are familiar, at any rate in the abstract, and we shall have occasion to return to them in discussing the problems of diagnosis. Unfortunately the limitations to the use of this method are such as to reduce very much its scientific value.

The chief conditions that must be secured may be thus enumerated: (1) The effect of the drug must be decided and unequivocal. (2) Its influence must not be obscured by any other change in the measures employed, which might cause the improvement observed. (3) The lesion must not be one of which the symptoms tend to spontaneous subsidence. (4) The lesion must not be one that can be influenced by the drug, when it is not due to syphilis. (5) Before a negative result can be admitted as evidence that the lesion is not due to syphilis, we must be satisfied that the damage produced is not so great that its manifestations *cannot* speedily subside. These conditions render the process almost useless for our present object. The method is applicable chiefly to chronic lesions, in which the influence of treatment can be clear and decided. Acute lesions, if they are slight, have too great tendency to subside, and, if they are severe, have effects of too prolonged duration.

We may briefly glance at the lesions to which these methods of proof apply. First, there are two important acute inflammations—those of the membranes of the brain and of the substance of the spinal cord. Are these ever due to syphilis? Unfortunately, I can only ask the question, and try to show how to obtain the answer, which I am not able to give. That syphilis causes chronic meningitis is one of the surest facts of medicine. It causes chronic inflammation of the membranes of the brain and of the spinal cord, and either may be diffuse or focal. The latter is usually in the pia mater, and is associated with the formation of caseating new tissue. In the cord it usually begins at the surface and involves the nerve roots, so as to give rise to obtrusive radiating pains, or muscular wasting. The diffuse inflammation more often involves the dura mater and is attended with the production of fibrous tissue constituting in the cord what has been called "hypertrophic pachymeningitis," which compresses the cord itself and gives rise to grave symptoms. Within the skull bilateral headache and general convulsions may be its expression, whereas the focal inflammation gives rise to fits that begin locally. These facts are very important in respect to diagnosis. Whether syphilis also causes acute meningitis, such as reaches a considerable degree of intensity in a few days, is still uncertain. Analogy suggests that an acute meningitis, like an acute iritis, may be due to it, but this is enough only to make us search for evidence. Morbid anatomy, as we have seen, is not likely to supply an answer. The therapeutic test is of little value, on account of its essential limitations, for we dare not secure the absence of other influences; we are bound to do all we can for the safety of the patient, and there is reason to believe that mercury may determine the subsidence of an inflammation that is not syphilitic. I have once known the symptoms of a general meningitis to develop most acutely in a man who had suffered from syphilis, and then to slowly pass away under the influence of mercury; but such a case can alone do no more than raise or strengthen a suspicion; it cannot supply the needed proof. Acute meningitis is very rare in adult men, but it does occur when we can exclude syphilis, and hence the question of causation can be decided only by a series of cases. The readiness with which acute meningitis is ascribed to syphilis by some writers is astonishing, or would be astonishing had not the capacity for wonder been already diminished by the process of the diagnosis of the disease. To quote one instance: A patient suffered, for a few days only, from pain in one temple, an elevation of temperature of half a degree Fahrenheit, and diminished frequency of the pulse. This was considered to justify the diagnosis, not indeed of meningitis, but of a lesion safe in the region of hypothesis, local congestion of the membranes. This assumed lesion was further ascribed to syphilis, because the symptoms occurred twenty-five days after—not definite constitutional symptoms, not even after enlargement of the glands, not even after the primary induration was discovered, but twenty-five days after the exposure to infection! We scarcely marvel after this to find that the case is regarded as affording strong support to a theory —indeed, as part of its proof—that the early symptoms of syphilis frequently include congestive lesions of internal organs, especially of the nerve centres, which are analogous to the early eruptions on the skin. I think our time may be better occupied than by discussing this theory in detail.

The possible relation of myelitis to syphilis is one of the most difficult problems connected with this subject, and it is one on which definite evidence is specially needed. The term "myelitis," even in its narrower sense, includes diseases widely different in their nature and pathological relations, and these relations are still imperfectly known. We can, however, distinguish acute transverse myelitis, sub-acute disseminated myelitis, and acute poliomyelitis, as fairly well-defined forms, which have distinctly different causation, and the relation of each to syphilis must be separately ascertained. In each form this investigation has still to be made. The fact that special observations are necessary renders it difficult for the evidence to be obtained by any one person. It is certain that not one of these forms is exclusively related to syphilis, and it is not probable that the relation is, in any one of them, so preponderant that the fact could be decided by a small number of observations, for we must depend almost entirely on the evidence of sequence. The limitations to the method of therapeutics are here prohibitory. The course of the symptoms is no more guide to that of the inflammatory process than is the course of a cardiac murmur evidence of the course of the endocarditis to which it was due. If the inflammation of the spinal cord is trifling, the tendency to spontaneous improvement deprives the subsidence of the symptoms of any significance. Moreover, there are some facts that suggest that the less acute forms may be due to syphilis, and yet may not yield to its treatment. In the case of an inflammation such a statement seems self-contradictory, but in dealing with facts we must lay aside all our preconceptions. I mentioned just now a form of *insular* myelitis, possibly always syphilitic in origin and specific in its distribution. (To avoid confusion I may point out that in this form the inflammation is in small islets; in the common form of disseminated inflammation larger scattered tracts or areas are affected.) If the insular myelitis is syphilitic, it does not seem to be distinctly amenable to treatment. In the cases hitherto observed the patients have died. In some of them, as in my own, antisyphilitic treatment had no influence on the disease.

I have seen several other cases of sub-acute and sub-chronic myelitis in syphilitic subjects, in which neither iodide in large doses, nor mercury pushed to salivation had any appreciable influence on the lesion, either on its extension in the cord or its intensification in the parts first affected. This seems also to have been true of the remarkable case of chronic caseating inflammation described by Charcot and Gombault, to which I have already alluded as a lesion pathologically specific.

Here, then, we are driven back on sequence for our evidence of causation, and the needful observations, serial and unselected, have yet to be made. The same statement must be made regarding transverse myelitis, the most common variety, recognizable during life without much difficulty. My own experience is suggestive only. I have seen several cases in which the disease followed syphilis, more, I think, than mere coincidence will explain. On the other hand, I have met with no cases of acute poliomyelitis suggesting a syphilitic causation, and the significance of a few cases that have been published as evidence of such causation seems to me open to doubt. We must remember that the sub-acute form has only lately been distinguished from peripheral neuritis, and the most common cause of peripheral neuritis, alcoholism, is sometimes associated with syphilis as a consequence of mode of life.

Another acute affection of the spinal cord, the mysterious acute ascending paralysis, has also been supposed to be a result of syphilis, partly on the ground of sequence, partly on that of recovery under the influence of mercury. The proportion of published cases in syphilitic subjects is certainly large, and justifies careful investigation. In considering the influence of mercury we must remember that there is some evidence that other acute toxic states besides that of syphilis are modified by mercury. In the course of some investigations made in 1885, at the Brown Institution, Dr. Cash discovered that animals, brought under the influence of mercury, resist otherwise fatal inoculation with the anthrax bacilli—a remarkable fact, which does not seem to have attracted the attention it deserves. We must also remember that the syphilitic organisms may leave behind them a product of a different nature which may not be amenable to the influences which affect the organisms themselves.

Regarding the focal sclerosis of the nerve centres, including what may be termed "chronic sclerotic inflammation," I have little to say. The evidence seems to be against any relation to syphilis of true insular sclerosis. Only future observations, for which we may have long to wait, can determine the question as regards certain other extremely rare forms of diffuse sclerosis. A remarkable case of miliary degeneration of the gray matter of the brain, published by me some years ago, occurred in a syphilitic subject. Many years ago I examined for Dr. Hughlings Jackson a brain of an old syphilitic, which presented a remarkable induration of a large part of the white substance of one hemisphere, without change in tint. (The history of this case, unfortunately, we have been unable to trace.) Lastly, sclerosis of the cortex in infantile syphilis, as we have already seen, may possibly be sometimes primary, although generally it is the result of inflammation in infarcted areas.

The processes we have hitherto considered originate either outside the nerve centres themselves, or, when within the nerve tissue, are essentially interstitial. Of all, therefore, it seems to be true that the primary morbid process is independent of the nerve elements themselves. These elements suffer just as they would from any similar process, not syphilitic in nature. It is on these simple processes in the nerve elements, and on these only, that most of the symptoms depend—a fact of the utmost practical importance.

The spinal nerves seem to escape the influence of syphilis, and the cranial nerves suffer chiefly at the base of the brain, where several are involved in a focus of inflammation. To this some of the nerves to the eyeball seem to offer an exception; they may suffer alone. The facial nerves present a remarkable freedom. The nerves that arise from the side of the medulla are often involved, especially the hypoglossal and spinal accessory, causing palsy of the tongue, palate, and vocal cord, on one side. These palsies are seldom due to any other cause, so that their occurrence should suggest syphilis at once to your mind.

Paralysis in the arm, due to deep cellulitis in the neck, chronic and indurating, is another occasional result. Note, however, that, as a rule, the spinal nerves, if they suffer, do so only on one side and in consequence of damage to their roots.

Until lately it was believed that the influence of syphilis was limited to the production of such lesions as we have hitherto considered. But modern observations, led by those of Fournier, have resulted in a considerable amount of evidence that the effects of syphilis are not thus limited—that the disease is followed by lesions of a totally different nature, and is followed by them in a way that admits of no other explanation than that it causes them. These lesions are degenerations of the nerve structures, with such a distribution according to function as can be explained only by assuming that the process begins in the nerve elements themselves. The relation of these "system degenerations" to syphilis was first asserted in the case of tabes, locomotor ataxy, and this disease is still the most conspicuous example, around which most of the controversy has turned. The assertion that syphilis causes lesions so totally unlike the known effects of the disease was sufficiently startling to be received with surprise and suspicion, a suspicion not lessened by the fact that the only evidence of the relation was, and still is, that which depends on numbers for its weight. Various objections were urged. It was said that syphilis could not cause such disease; that cases of true syphilitic lesions had been mistaken for degenerations; and that if tabes were due to syphilis the lesion must be primarily interstitial, and not a primary nerve-degeneration. The common absence of benefit from anti-luetic treatment was urged in support of the objection, and the statistical evidence was met by other figures of different significance, obtained from existing records. Some of these arguments, however, were seen to be untenable as soon as they were clearly stated. The assertion that syphilis *could* not cause a lesion different in nature from the effects previously known, was too obviously a transfer to medicine of the method employed by Canute upon the sea-shore. The absence of evidence of one kind, that of therapeutics, was manifestly no refutation of a totally different species of proof. The comparison of statistics, moreover, resulted in a curious and instructive episode in the history of medicine. When those observers who had gathered their cases from old records came to make fresh observations on the subject, they found that the two sets did not correspond. They found that the proportion of syphilis yielded by the special observations was much higher, and in more than one instance the special inquiry transformed an opponent into a supporter of the new opinion. It was one more instance of the fact I just now mentioned—the uselessness of figures based on observations not made with the special question in view. .

The proportion of cases of tabes in which syphilis has been traced by different investigators has varied much—from 90 per cent. down to a proportion that has been held to be not far above the range of accidental association. Instead of detailing to you the various statements that have been made, it may be more instructive to give the figures supplied by a consecutive and unselected series of cases that has come under my own observation. Of 170 cases of characteristic tabes in males there was evidence of syphilis (in a few confined to a chancre certainly hard) in 55 per cent. In another 14 per cent. there was a history of a sore only, of unknown nature. In only 31 per cent. could neither be ascertained. Of the 170, 92 were seen in private, and 78 were hospital patients. The percentage of syphilis in private cases was 57, in the hospital patients 52. Probably the difference is due to the more accurate information to be obtained from those who are most likely to notice and remember such diseases. It is remarkable, however, that the proportion between the cases of syphilis and the cases without even a chancre is exactly the same in the two sets, as 11 to 6. The difference in the percentage arises from the fact that a history of a sore of unknown nature was almost twice as frequent among the hospital as among the private patients. In either series the addition to the syphilitics of one quarter of the cases of chancre will give about 60 per cent., and a deduction for mere accidental coincidence would have at least 55 per cent. of unquestionable causation. But it is certain that to this a large addition must be made for untraceable syphilis. These figures are obtained by placing all doubtful cases on the negative side, including many with a strong presumption of syphilis. If it is true that latent syphilis is common, and if the disease can be proved to be a cause of tabes in more than half the cases, is it not almost certain that a large proportion of the remainder are really due to it? In most of them no other cause could be traced, and in not more than 5 per cent. of the whole had there been no exposure to the risk of infection. Although the *method* which discloses syphilis in an exceedingly large proportion of the cases may be open to question, I cannot but think that the observers who describe it in 75 or 80 per cent. are not far from the real truth.

Not only is the relation of syphilis to tabes beyond reasonable doubt, but we are beginning to discern that there are other diseases that stand in a similar, although unequal, relation to the disease. One other group of degenerations we can, indeed, connect with syphilis as surely and as extensively as tabes—the degenerative ocular palsies, which, often associated with tabes, occur also alone. There is strong evidence of a similar relation in simple atrophy of the optic nerve. Among the motor palsies, one of the most important to the physician, although the least important to the patient, is the isolated loss of the light-reflex of the iris, with which the name of Argyll-Robertson is connected. I have notes of only twenty-two cases in which this symptom existed apart from other degenerative affections of the nervous system, but I believe that it is much less rare than this number might suggest. It is seldom looked for, unless other symptoms suggest degenerative disease. Of these twenty-two cases, syphilis was certain in twelve, and highly probable in three others, while two more had a history of a venereal sore. In six cases there was no history or evidence of either syphilis or a chancre; nevertheless, two of them had had an attack of sudden hemiplegia in early adult life, without any indication of a cause of embolism—an event that always affords ground for suspecting syphilis. From these facts we may feel confident that syphilis existed in two-thirds of the cases, and we may suspect that it existed in three-quarters. One of the cases was a young girl, the subject of inherited syphilis, and another case was this girl's mother.

The relation of this symptom to syphilis makes it one of great practical value. It proves that a degenerative process is at work in the nervous system, and it raises a strong presumption that this process is the result of syphilis. It is true, the dependence on central degeneration has not been proved by observation, but the indirect evidence of it is very strong.

The more extensive palsies, external and internal, to which the term "ophthalmoplegia" has been applied are much more rare. They occur in syphilitic subjects in association with tabes, and also alone, as was clearly proved by Mr. Jonathan Hutchinson. In the case of external ophthalmoplegia, in which I made an examination of the nerve centres, preceding syphilis was certain (the evidence I have already mentioned), and the lesion was as pure a nerve-degeneration as can be conceived. But these diseases, like tabes, are not exclusively related to syphilis; they certainly occur, now and then, in persons in whom syphilis can be excluded.

This evidence seems to leave no room for doubt. It receives, moreover, a strong emphasis from the remarkable fact that nerve-degenerations such as succeed acquired syphilis in the adult, are met with also in young persons who are the subjects of the inherited disease. Such cases, it is true, are rare, but their rarity may be due, less to any difference in the influence of the disease, than to the greater vital stability possessed by younger tissues. I have met with several examples, and their interest may justify a brief mention of some. One case was a lad who, in infancy and early childhood, suffered much from various manifestations of syphilis, including slight hemiplegia and choroiditis. At the age of seventeen the symptoms of locomotor ataxy were distinct. There was unsteadiness in standing and walking, there had been lightning-pains, and no trace of the knee-jerk could be obtained. It was difficult to ascertain when the symptoms commenced; the unsteadiness had only attracted attention during a year or so. A second case was that of a girl aged fifteen (seen through the courtesy of Mr. Nettleship). She had typical teeth, and traces of characteristic inflammation in cornea and choroid. The left knee-jerk was absolutely lost, and only a very slight movement could be obtained on the right side. There was no ataxy or anæsthesia. In each iris the light reflex was quite absent, although there was full action to accommodation, and vision in one eye was normal. In another case the disease was thought to be pseudo-hypertrophic paralysis because there was no knee-jerk and the child rose from the floor with difficulty.

Another malady of great importance in regard to which this question has arisen is general paralysis of the insane. The disease is probably to be classed with primary nerve-degenerations, but in many cases, it would seem, the secondary changes attain a very considerable independence of degree. Its close alliance with tabes is shown by the conditions of age and sex in which it occurs, by the degenerative ocular palsies common to the two, and by their actual combination in many instances. This would lead us to expect a similar relation to syphilis. The scientific proof, however, is beset with many difficulties. The statistics obtained have varied greatly, but no less than 75 per cent. of certain syphilis is described by Mendel in 146 cases. A proportion almost as large has been met with by others. One serious difficulty arises from the great variations presented by the disease. Typical cases form a minority of those that may be classed under the general name, and both the typical and untypical forms certainly may occur independently of syphilis. I believe, however, that syphilis can be traced in a large proportion of each distinguishable variety, although I am not able to offer you any definite figures. Fournier has pointed out how often symptoms resembling more or less closely those of general paralysis seem to result from organic syphilitic lesions of the brain, and I believe that the explanation of this is to be found, less in the fact that the organic diseases cause these symptoms, than in the profound degenerative tendency which so often results from syphilis. This tendency may accompany the organic lesion, and is perhaps sometimes actually excited by it. I have occasionally observed that degenerative symptoms, mental and articulatory, may distinctly come on soon *after* a sudden organic syphilitic lesion, and may subsequently lessen, and even pass away.

Another central degeneration that is probably an occasional consequence of syphilis is chronic muscular atrophy, due to degeneration of the motor cells, with or without the signs of lateral sclerosis. Great care is necessary to avoid the error of mistaking for this the atrophy that results from syphilitic pachymeningitis. In several cases that I have seen, the true degenerative affection has been an early sequel of syphilis, and no other cause could be traced. Similar cases have been published by others, but they do no more than *suggest* the relation, although they suggest it strongly, and my own notes do not enable me to offer any statistical evidence on the question. The difficulty of the investigation is increased by the large number of cases of senile character. The exclusion of these would simplify, without hindering, the search for evidence. The investigation should thus be restricted to cases commencing between twenty-five and fifty, the standard being limited to the same age.

The other common degenerative diseases of the spinal cord are associated with syphilis only, in a proportion so small as to raise a doubt whether there is any causal relation between the two. Nevertheless, they occasionally follow syphilis when no other cause can be traced.

This is especially true of primary lateral sclerosis, that probably often underlies spastic paralysis. Ataxic paraplegia, the combined effect of lateral and posterior sclerosis, seems to be very seldom preceded by syphilis.

It must be admitted that the relation of these degenerative diseases to syphilis is a startling extension of our knowledge of its influence. And yet, perhaps, our surprise is not altogether justified. Other facts exist which, if they do not lessen the marvel, at least may prepare us for it. We are familiar with the influence of chemical poisons on certain parts of the nervous system,—with the way in which atropine, curara, digitalis, and strychnia select for isolated derangement certain nerve elements, leaving adjacent structures unaffected. This must be the result, not only of the nature of the poison, but also of the nature of the nerve elements influenced, through which they, and they only, have a responsive susceptibility. That an organized virus may exert a similar elective influence, the symptoms of hydrophobia clearly show, and those of whooping cough at least suggest. Further, it is, I think, most important to note that the phenomena and pathology of diphtheritic paralysis demonstrate that an acute degeneration of certain nerve elements may also be the effect of an organic virus. We may note, also, that the effects of the poison of diphtheria sometimes so closely resemble those of the poison of syphilis, that cases of diphtheritic paralysis have been repeatedly mistaken for tabes, and we have seen that syphilis has, like diphtheria, a special tendency to derange an intraocular muscle. Still further, in the circumstance that the palsy is a sequel rather than a concomitant of diphtheria, we have a fact that is most significant, and a fact whose significance is not destroyed by the difference of interval in the two cases, since the difference of interval is not proportionately greater than is the difference in duration of the primary maladies. Diphtheria runs its course in a few days, and its effects on the nervous system are developed in weeks or months. Syphilis lasts, in the stage of a blood disease, at least for months, and perhaps for much longer, and its effects on the nervous system are developed only in the course of years. But there is possibly even more than an analogy between the effects of the two diseases. I have seen a few cases in which there was evidence of a persistent, and even a progressive, lesion of the spinal cord after diphtheria, and I have seen three cases in whom true primary atrophy of the optic nerves, with partial external ophthalmoplegia, was a distinct degenerative sequel to diphtheria.

It may, I believe, be regarded as a general law, and a law of much importance in practical diagnosis, that the isolated impairment of nerve structures that have a certain function, when acute, indicates a toxic influence —when chronic, a degenerative process. It is instructive to note that these two mechanisms are not entirely dissociated—that the degenerative process may be the late result of a toxic agent.*

The problems we have last considered have an interest which is at present, alas, purely scientific. It is a strange anomaly, this scourge of the sinister side of civilized life. Its direct effects we can control more than those of any other malady of like nature, and yet both the essential element of the disease, and its remote

---

* It has, indeed, become highly probable that they are due to a toxic agent that is produced by the syphilitic organisms, and is left by them in the system—an agent that may be a chemical material and not an "organized virus." This was suggested by Strümpell. (Neur. Cent., 1889, p. 547.) The theory agrees with the fact that other acute specific diseases, more acute in course, seem to have a like effect, since they may be followed, even after some months, by degenerative multiple neuritis. (See "Manual of Diseases of the Nervous System," vol. i.) This theory enables us to understand why the treatment that is so powerful over the true syphilitic has no influence on their degenerative states —a point to which we shall return. We can also understand the bilateral symmetry of the symptoms, so different from the irregular distribution of the true syphilitic lesions. It is assumed that the latter are due to the fixation of germs and their development in the tissues, while the degenerations result from the influence on the nerve elements of the toxic agent circulating in the blood reaching all parts and affecting those that possess a similar susceptibility, which, of course, corresponds on the two sides.

The hypothesis was suggested by Strümpell before the delivery of these lectures, but had not then come to my knowledge.

effects, seem alike beyond our reach. But the knowledge that is not power now may be power in the future. When the specific organism has been identified and isolated—when it has yielded the secret of its life history—then we may look for means, if not of destroying it, at least of modifying its processes and effects, and of modifying them not only in the experimenter's flask, but within the human body. Until this can be accomplished, the light of research does but reveal, in vaster range and more complete detail, the grim features of the malady. Still, the light is welcome. We see how far the grasp of the disease extends into regions believed to be beyond its reach. We see how long its consequences endure, and that they live on after the disease itself has ceased to be a power for evil, but we see also how much more vast than was formerly conceived will be the saving influence of prevention or of cure, when the knowledge that is power is in our hand.

LECTURE II.

Mr. President and Gentlemen,—From the questions of pathology and causal evidence that occupied us in the last lecture we pass to-night to subjects of more immediate practical interest—those of symptoms and diagnosis. The transition may perhaps bring with it a sense of relief, but, unfortunately, it brings with it no escape from the difficulties in which the subject is involved, and which surround, not less closely and scarcely less consciously, its practical aspects.

First, however, I should supply one omission in the last lecture. To the long list of maladies, the certain or supposed consequences of syphilis, should be added some of the functional diseases of the nervous system. By some authorities it is thought that such diseases as epilepsy, hysteria, insanity, and neuralgia, in their purely functional form, are frequent consequences of the influence of the virus on the nerve elements during the period of its greatest activity. Time fails me for any discussion of the subject, but there are two points that are especially important in any examination of the subject. The first is that if any special forms of functional disease seem to be exclusively related to syphilis, it is essential that the positive evidence from the side of syphilis should be supplemented and confirmed by negative evidence that these special and peculiar forms do not occur apart from syphilis. Secondly, if the forms associated with syphilis are not special, it is needful to consider most carefully the question of indirect causation. The causes of these diseases are many and various, and they are widely prevalent. Some of the most potent are involved in the conditions and influences inseparable from constitutional syphilis and its treatment, in which physical depression and mental annoyance are sometimes combined with a reaction from a life of intense excitement of body and mind. It is also necessary to ascertain and allow for every traceable predisposition, and likewise to see that the therapeutic evidence is secured from fallacy by a rigid observance of the conditions that were mentioned in the last lecture as being essential. I cannot help thinking that these measures would alter considerably the aspect of much of the evidence on which the opinions rest, and would remove the uncertain to an extent that would reduce the certain to its vanishing point.

In the diagnostic study of symptoms we still feel the pressure of the antagonism I mentioned at the outset. The success, incomplete though it be, with which we are able to treat syphilitic diseases of the nervous system, renders it impossible to obtain a knowledge of their symptoms as extensive or as certain as we have gained in the case of most other maladies. We can attain certainty, or at least the high probability, which in medicine has generally to do duty for certainty, only by confining our attention to cases in which post-mortem evidence places the nature of the disease beyond doubt. But if we thus restrict our attention to these certain facts, and our inferences to the sure conclusions we can draw from those facts, our knowledge will be at once small in extent and imperfect in character. It will be based only or chiefly on cases of great severity, and the effects of slighter forms and degrees of disease will, to a large extent, escape us. If we include the latter in our investigation, we are upon ground which is always insecure, and often treacherous; for an uncertainty, varying with the knowledge and care of the observer, will attach to a large number of our conclusions. This uncertainty ought to be frankly admitted. If we recognize its existence, if we employ the probable conclusions as such, and do not impute to them the certainty they lack, we may use them, or many of them, without much risk of being misled. To obtain the greatest practical power, we must, on the one hand, be often content with probabilities that are not high, but, on the other hand, we must use these probabilities as such, and trust them only so far as their nature warrants. Thus we may obtain knowledge of the highest value, in which we can place implicit confidence.

In the practical diagnosis of these diseases, just as in their scientific investigation, it is necessary to weigh carefully each step in the reasoning, and to do so far more carefully than in the case of many other maladies. Remember that the amount of conscious and deliberate attention to the process of inference, which it is desirable to give, varies much in different kinds of disease. It should always be reduced to the least amount that is consistent with safety, but no further. So far, however, we must go. Did we weigh with care each step in every diagnostic act, the limits of patience and of life would be reached before a quarter of the work had been done that might otherwise be accomplished. Here, as so often, our proceeding has to be based on a compromise between the ideal and the adequate. Unconscious inference enters, of course, largely into the simplest so-called "observations," and into such diagnosis as seems to consist of pure observation. Still larger is the share it takes in the recognition of diseases that are inaccessible to direct observation, and yet yield "physical signs" of their presence. But, in the latter case, we can allow it to remain unconscious without much risk of error, and we habitually do so, and even speak of "finding" a pneumonia or a pleurisy without a suspicion of error in our language, although we really infer the existence of the lesion by a series of mental processes that would be of considerable length if the steps were distinguished. Our short cuts, in such cases, involve little danger to an observer who has been properly trained. But the lesions in the nervous system are wholly beyond the range of observation, however aided, and with all our looseness in the use of words, we do not speak of "finding" a cerebral hemorrhage, or even a cerebral tumor, during life. In purely inferential diagnosis, in which the disease manifests itself only by effects that are not only indirect, and, indeed, often doubly indirect—the effects of effects of .the lesion—we cannot dispense with the careful consideration of each step we take. This statement is eminently true of the diseases we are now considering, because in them we have to deal not only with the character of the lesion, but also with its nature.

The fact indicated in the last words is the first important consideration in regard to the nature of the diagnostic process. The process is, and must be, always a double one. Of the two parts, one should always be taken before the other. We should always ask what is the seat and what is the nature of the morbid process, before we attempt to decide whether or not it is specific and due to syphilis. In greater detail, the essential steps are these: (1) What is the seat of the lesion, as indicated by the symptoms, and what is its nature as shown by their course? (2) Is the process thus indicated one of those that *may* be syphilitic? (3) Has the patient had syphilis? (4) Can any other cause of such a morbid process be traced? (5) Lastly, and subsequently, we have to see whether the result of treatment confirms our conclusion. The last element is different from the rest in character and separate in time. Its consideration involves questions that will come under our notice in the next lecture, and, moreover, are those we examined in considering the scientific use of the therapeutic test. They need not, therefore, detain us to-day. But the other steps we must consider in the abstract, because the only mode of avoiding error in these cases, and of reaching as sure a conclusion as the conditions permit, is to treat each case as a new problem, irrespective of other cases or generic types, and to work it out by taking each step separately—comparing the steps only at the end of the process.

One of these questions—whether any other cause can be traced—may seem superfluous, when we are dealing with processes that have no other cause than syphilis. But this statement of "no other cause" is true only of the pathological lesion. Most syphilitic processes have their analogues in processes that are not syphilitic. The effects of the two on the nerve elements may be the same, and the symptoms due to those effects may be the same. A gumma of the cortex, and a glioma in the same part, may produce identical symptoms. The consequences of the occlusion of an artery by syphilitic disease and by embolism may absolutely correspond. The element of specificity, therefore, is absent during life. There are no symptoms of which we can say, as we can of post-mortem aspect, "this proves a specific lesion." There are no symptoms, and no combinations of symptoms, produced by syphilis, that are not also produced by other causes.

While all specific processes have their simple analogues, there are many simple processes that have no specific analogues, and many of these are manifested by characteristic symptoms. Some of these simple processes may be due to syphilis, as we have seen, but even then they are not specific; so far as we can discern, they are simple. If they are due to syphilis, they are certainly not influenced by treatment as the specific lesions are. Thus, while there are no symptoms that prove a specific lesion, there are many symptoms that prove that a lesion is not pathologically specific.

Hence arises the paramount importance of considering, in the first instance, what kind of lesion the symptoms indicate. To neglect this question, to conclude that because a patient has a disease of the nervous system, and has had syphilis, the disease is syphilitic, is a proceeding precisely the same as if we assume that a sore throat is syphilitic without looking at it. If a syphilitic person has an attack of hæmatemesis, it would not be considered quite logical to assume at once that the hæmatemesis was of syphilitic origin, and yet observed facts actually afford more justification for this opinion than for the conclusion that paraplegia of absolutely sudden onset is due to syphilis. We know that syphilis has caused hemorrhage into the stomach, but we are not certain that it has caused hemorrhage into the spinal cord.

Thus a consideration of the kind of lesion that must exist to cause the symptoms may enable us to say one of two things—either that the process is one that may be specific, or that it is one that cannot be specific. To deal with these indications in detail would be to enter into the whole region of the diagnosis of diseases of the nervous system. There are, however, two conditions to which almost all specific lesions conform, one of time, the other of place. First, true specific lesions are generally either sudden, or sub-acute, or sub-chronic. An acute onset—acute, not actually sudden—practically means acute inflammation, and we saw in the first lecture that the occurrence of acute syphilitic inflammation must be considered as still *sub judice*. As a matter of fact, the symptoms of certainly specific lesions very seldom develop to a considerable degree in less than a week. They are also very seldom actually chronic—seldom occupy more than three months in development. An actually sudden onset, however, is not at all rare; it is the common onset of the symptoms due to vascular occlusion. With these exceptions, when symptoms develop in a few days or many months, they are not likely to be due to a true specific lesion. The degenerative sequelæ are, of course, chronic in onset, but they are not truly specific. Moreover, the chronic interstitial inflammations, doubtfully specific, are also chronic.

Secondly, as to place. The specific processes are, as we have seen, outside the nerve-elements, and they have therefore no special relation to the nerve-functions. Hence their effects are random in distribution. They are related to a special function only when that function is subserved by one region, and this is so rare as to be of little practical importance. As a general rule, if we find certain structures of common function selected, for isolated impairment from among others of different function, we may be sure that we have not to deal with a true specific process. It may be a post-syphilitic degeneration, but it is not a really specific lesion. These two indications are thus chiefly of importance as enabling us to exclude processes that are not syphilitic.

Assuming that we have evidence of a process that *may* be specific, we can often go a step further, and say that, from the symptoms, it is *likely* to be specific. But before doing this, it is desirable to ascertain whether, in the case before us, this possible cause can be traced The lesion may be due to syphilis—is this particular cause an actual antecedent in the case before us? In every department of diagnosis, the fact that a certain cause can be traced may always justly be allowed weight. It must never take precedence of more direct evidence of the nature of the lesion; but in due subordination, it is an indication of great value. Hence the next question is, has the patient had syphilis? At least, this is the form in which the question is commonly put; and in many cases this form is sufficient. But we must not be content with a negative answer. We have already seen that some of those who have had syphilis do not know it. We cannot absolutely exclude syphilis unless we can exclude infection. The disease is so seldom contracted except in one way, that we need not take other possibilities into consideration. Hence, if there has been any possibility of exposure to infection, the disease cannot be considered to be out of the question. The truth and importance of this will be doubted by no one who has seen much of the late effects of syphilis, but to those under whose notice the disease seldom comes, it may seem scarcely credible. Indeed, as a matter of fact, few statements are more often doubted.

In connection with this point, the question arises,— presuming the disease to be possible, what influence should uncertainty as to the fact of syphilis have on the diagnosis? How far should probability of the syphilitic nature of the lesion be diminished by the improbability of past syphilis, and *vice versa?* I think that the common error is to allow too much weight to this element of uncertainty, but it is an error that is not easily avoided. The symptoms, in a given case, are of such a character that their cause may be a syphilitic process; if the patient is known to have had syphilis, the syphilitic nature of the process seems more probable than if his syphilis is doubtful. It is right to attach some weight to the point, but the weight should be attached to the element in the diagnosis to which it properly belongs—the element of probability. It must not be allowed to influence any distinct indications of either a positive or negative character. If, for instance, the symptoms are such as distinctly indicate a syphilitic lesion, a lesion scarcely ever due to any other cause, it is evident that this indication is not really influenced by the fact that the evidence of syphilis is inconclusive, provided the disease is possible. On the other hand, if the symptoms are such as to make it most unlikely that the process is syphilitic, this improbability is not materially lessened, even by a certainty that the patient has had the disease.

If the lesion is one that may be syphilitic—if previous syphilis is certain or possible—the question arises, Can any other cause of such a morbid process be traced? If any other cause exists, the difficulty is greatly increased. The diagnosis then depends on a comparison of the precise character, first, of the lesions produced by the two causes with that indicated by the symptoms, secondly, of the symptoms by which the lesions are manifested with those presented by the patient.

This abstract outline of the process of diagnosis may seem to you needlessly complex. You may think that these considerations are purely theoretical, but they actually represent, in the simplest form, the essential processes adopted in actual diagnosis. I give them in the abstract, because their application varies in each case, and for their effective use familiarity with their principles is essential.

It must, however, be frankly admitted that exact diagnosis of the nature of these specific lesions is sometimes impossible. The symptoms may be equivocal, such as may be produced by more than one syphilitic process, but we can generally narrow the probabilities to two lesions.

We may now consider the symptoms of a few of these specific processes. A syphilitic gumma causes, for the most part, symptoms like those of any other intracranial tumor of rapid growth—general cerebral symptoms, and focal symptoms due to the local influence of the growth. The special features are never more than suggestive. They depend on the course, and also on the fact that gummata are rather more frequent in certain parts than in others. The course, like that of other specific lesions, is sub-acute or sub-chronic; the seat is generally superficial, in the cortex or at the base, sometimes within the pons, seldom in the cerebellum, occasionally in the thalamic region, growing in from the side of the crus. Hence symptoms of cortical irritation are relatively more frequent than in other tumors, and convulsions are common. They give most important information from their local distribution or commencement, or from the aura that equally indicates the spot at which the discharge begins. But all these indications, even when combined, furnish no absolute or even approximate ground of distinction. They may be, and, indeed, often are, produced by growths of other nature, especially by gliomata of rapid growth.

The indication of the course of the lesion afforded by the symptoms is chiefly useful for its negative significance; a very chronic growth is not likely to be syphilitic. On this point we may gain instructive information from the optic neuritis that is so often present, the course of which generally shows the course of the disease of the brain. Syphilomata always cause an acute form of optic neuritis, becoming intense. A rapid growth never causes a chronic form of neuritis, although now and then a slow growth may cause an acute form. Hence, while acuteness of the neuritis is of little diagnostic value, chronicity—a neuritis that remains for a long time moderate or slight in degree—is distinctly opposed to the diagnosis of a syphilitic growth, and adds considerable weight to the similar indication afforded by great chronicity of other symptoms. This indication is especially valuable when the early symptoms are equivocal, and we find it difficult to say how long the tumor has existed.

These growths are among the specific lesions in

which the effect of treatment is of most diagnostic value. It usually causes a prompt diminution in the symptoms, but, to be of significance, this diminution should be considerable, and should involve both sets of symptoms, the local and general. It is a remarkable but certain fact, that now and then a considerable diminution in symptoms of other kinds of tumors, and especially in the general symptoms, sometimes follows the administration of potassium iodide. The effect is most frequent in gliomata, and these are the growths that give rise to most difficulty. Sometimes even the optic neuritis may lessen, but more often it is unchanged, and this should put us on our guard. In estimating the significance of the subsidence of the neuritis, we must remember that every intense inflammation, after it has reached its height, slowly subsides, although the tumor that has caused it continues to grow. We must not mistake this natural subsidence for the result of our treatment. This therapeutic difficulty can only be met by waiting and watching. If any of the symptoms return or increase, if any symptoms persist that would be readily influenced were the growth syphilitic, we must assume that it is not likely to be of this nature. The delay cannot be avoided, but it is a grave disadvantage, because the progress of surgery has so greatly increased the need, not only for exact, but for prompt diagnosis.

The chronic local meningitis of syphilis causes distinctive symptoms chiefly when it affects the base of the brain and damages the cranial nerves, or when, at the convexity, it implicates the motor region. The symptoms are such as indicate a surface lesion, and the absence of the signs of any considerable loss of function of the subjacent tissue is the chief distinction from a gumma in the same situation. Especially when the symptoms indicate a wide area of irritation with merely superficial damage, the diagnosis can be made with considerable probability. This focal meningitis is probably less frequent at the convexity than are gummata, and it is certainly less frequent there than at the base of the brain. In the latter situation the most important symptoms are due to the damage to the cranial nerves, and the most important diagnostic points are well known. The therapeutic test is available only in recent cases of meningitis. When the development of the new tissue into fibres has set in, this seems to go on in spite of treatment, and the inevitable cicatricial contraction may perpetuate the damage produced in the more active stage. In a case of focal meningitis that has lasted for some months, although it may afterward be proved to be syphilitic in nature, the symptoms are generally influenced but little, and occasionally not at all, by anti-syphilitic treatment, however energetic. The other forms of meningitis, cerebral and spinal, interesting and important as they are, I am compelled to pass over.

The inadequacy of the clinical history of a syphilitic lesion, when that history is based only on certain evidence, is conspicuous in the case of the disease of the walls of the arteries. The classical researches of Heubner, who first placed our knowledge on this subject on a secure footing, were based, of necessity to a large extent, on fatal cases. The clinical picture, delineated from those facts, faithful as it is so far as regards the effects of extensive and severe disease, is yet imperfect as a representation of the slighter and much more common consequences of the morbid process. Passing over the former, because it is well known by description, although now seldom met with, I would ask your attention to some facts regarding the slight forms. The importance of the early diagnosis of this condition is unequaled even among syphilitic lesions. There is no luetic process that is so likely to cause grave and lasting damage, no process the effects of which, when once produced, are so entirely beyond our power, and yet can be so surely averted by prompt treatment. For this reason, I think it is wise to devote more time to this than to the other special questions, and also because it is a disease in which the diagnosis during life can generally be made with sufficient confidence to give a high value to clinical facts.

The chief mechanism by which this process is effective in causing symptoms is, as we have seen, by the sudden closure of an artery or of a branch going off from the diseased spot. The suddenness of the closure is no doubt due to the fact that the final occlusion is effected by thrombosis, and in this respect the syphilitic disease does not differ from the other common form of arterial disease, atheroma. The closure has the usual consequence, sudden anæmia and necrosis of the brain tissue, unless there be sufficient anastomoses to permit a collateral circulation. This may be either complete, restoring function, or incomplete, so that the continuity of the tissue is preserved only by a process which results in sclerotic induration.

The characteristic symptoms are those of a sudden focal lesion of the brain. Until the degenerative period of life is reached, such a sudden lesion is excessively rare except from embolism. Embolism can usually be excluded if there is no valvular heart disease or other source of embolism, and if the patient has not recently suffered from a disease known to cause endocarditis. Hence it follows that the occurrence of such symptoms in an adult under forty-five, who has had syphilis, and has not heart disease, may generally be held to indicate syphilitic disease of a cerebral artery, and this diagnosis may be made with a confidence possible in few other syphilitic lesions. This confidence is confirmed by the occasional opportunities for a verification. When, under such conditions, the diagnosis has been made, and the patient has subsequently died, the arterial disease has been found, invariably so far as my own experience and reading have gone. It is true that, in rare cases, certain other causes of arterial closure must also be excluded. I will mention these presently; they do not really lessen the validity of the statement just made, since the need for considering them is rare, and is generally obvious. But this confidence can exist only before the degenerative period of life. When this period arrives, other causes of arterial closure come into operation, causes which we cannot exclude with the readiness and confidence with which we can exclude embolism. This introduces an element of doubt into the diagnosis of cases over forty-five or fifty years of age. Hence, in endeavoring to ascertain the history of these cases from merely clinical observation, it is desirable to confine our attention to those in which the diagnosis can be made with most confidence. The disease does occur during the later period of life; we have to recognize the fact in our practical diagnosis. It may be that this renders some of our conclusions from the earlier cases imperfect, but we must accept this imperfection in order to obtain as much certainty as possible.

The following conclusions are based on a series of cases that have come under my own observation, excluding those over forty-five years of age. The cases are fifty in number, but on some points, owing to omissions, the precise number available for comparison is somewhat less.* The patients were over twenty-five, with the exception of one, aged twenty-one. Males were to females as three to one, a proportion that may not be far from the incidence of syphilis itself.

Many observers have noted how various is the interval between the primary disease and the cerebral lesion, and the variation is conspicuous in this series. In one quarter, the lesion occurred during the first two years after infection, and the remainder were distributed during the next twelve years (two, nine cases; three to five, seven cases; six to ten, nine cases; eleven to fifteen, eleven cases; sixteen to twenty, four cases). There was an interval of nineteen years in a man who had well-marked syphilis at eighteen, and at the age of

---

* I have seen many other cases since, but I have not had time to analyze them.

thirty-seven an attack of right hemiplegia, while a year later paraplegia came on in a sub-acute manner. Eighteen years intervened in a woman who, at sixteen, married a man of unsteady habits; her first two children were born dead; hemiplegia came at thirty-five, with an onset characteristic of arterial thrombosis. The shortest interval was apparently only three months, but the case does not seem otherwise open to doubt. A man, at the age of twenty-one, had for the first time a chancre—a distinct hard sore, followed by indolent enlargement of the glands. Three months after the appearance of the sore he had sudden hemiplegia, which slowly passed away, but was followed six months later by a second attack on the same side. In these cases, as, indeed, in all those compared, a careful search revealed no other cause. In the only other case in which the interval was less than a year the attack occurred six months after infection.

By far the most common effect of the lesion was hemiplegia, indicating disease of the middle cerebral artery, and probably the closure of a branch to the central ganglia. In only one case did convulsions at the onset suggest that the damage was limited to the cortex. (Simultaneous central softening necessarily prevents such special indication of a cortical lesion.) In three cases hemianopia suggested that the arterial disease was in the posterior cerebral. It is evident, however, that these cases present only some of the effects of vascular disease, and do not present at all its gravest consequences. Lesions may occur in parts of the brain where they cause no focal symptoms, and, on the other hand, disease of the basilar and vertebral arteries sufficient to cause definite symptoms is seldom survived. These grave effects, however, are well known, and I therefore pass them over. In probably nineteen cases out of twenty the Sylvian artery is that by which arterial disease causes symptoms, a fact easily explained, and of such disease this series presents a fair example.

The degree of the paralysis and its course present variations as great as, and similar to, those met with in embolism. It may be slight and transient, or severe and then often lasting, with only such recovery as occurs from compensation—a slow return of power in the leg, and some in the upper arm, the hand remaining powerless, or almost powerless, with late rigidity. I have a few times known two attacks of hemiplegia to occur, at intervals of months. Very rarely symptoms of some other syphilitic process coincide with the onset. It is only in such a case, in which there is reason to suspect a gumma, that I have seen optic neuritis. In several cases the patients have at some other time suffered from paraplegia, such as might be caused by a gumma pressing on the cord.

The onset occurred during sleep in a third of the cases, a proportion nearly corresponding to the time spent in the sleeping state. During the waking state the onset is seldom attended by loss of consciousness; it was so attended in only one-tenth of the cases—a fact of much importance. Giddiness and vomiting were occasional accompaniments. Of great significance also is the fact that the onset, in more than half the cases, was preceded by headache. The pain was usually great, and was either general or chiefly on the side of the subsequent lesion. It preceded the onset sometimes for only a few days or a week, often for several weeks, rarely for two or three months. It is apparently in some way due to the arterial disease itself. Occasionally, for a day or two before the onset, there was a slight tingling or other sensation in the side afterward paralyzed. Thus, one patient, after ten days of severe headache, had repeated attacks of tingling down the side during two days before onset. The attack itself was often quite sudden, but in some cases it was deliberate, or occurred in two stages. The following are instances of these modes of onset:—

1. The arm and leg became weak for a few minutes, then recovered, and half an hour later became suddenly powerless.

2. Slight weakness came on, and continued for six days before the sudden hemiplegia.

3. The leg became weak in the evening, and during the night the arm became powerless.

4. The patient woke up one morning with loss of power in the leg, and during the next two days the palsy spread gradually to the arm and face.

5. The arm became suddenly paralyzed, and six days later aphasia came on, with a convulsion, doubtless from the occlusion of a cortical branch.

The diagnosis of this disease chiefly depends on the causal indications, negative and positive, the absence of other causes, and the presence of syphilis. The evidence of therapeutics here fails us. The syphilitic process in the artery, and the secondary thrombus that results, are wholly separate from the simple process in the brain—the necrotic softening on which the symptoms directly depend. Treat the patient as you will; remove as speedily and as completely as possible the disease of the arterial wall; you cannot restore the circulation, nor can you avert the resulting necrosis of the brain tissue when once the blood has clotted in the vessel. Compare two series of cases of hemiplegia, one due to embolism, and the other to this disease, and you will find that their general course corresponds, however wise and thorough may have been the treatment adopted in the syphilitic series. In each series we meet with cases in which the palsy is slight or transient, but these are cases in which the motor structures have not been involved in the softening, and have only been interfered with for a time because the lesion was adjacent to them. In each series we find cases of severe paralysis, because the motor path was interrupted or the centres were destroyed. Hence, by your treatment you may save the patient from a fresh attack, or from an extension of the old mischief, but you cannot save him from destruction of tissue or avert its effects. Recognize any warning there may be—the attack may perhaps be prevented, but once the vessel is occluded, your treatment is powerless over the accomplished injury. Thus these are cases in which the result of treatment has no diagnostic significance, and if a positive result seems traceable we may be sure that the improvement is a coincidence and not a consequence. An illustration may make this important fact more definite. If the illustration is hypothetical it will nevertheless be felt, I think, to be within the bounds, not only of the possible, but of the probable. A man is seized with sudden hemiplegia due to some embolism, which has caused a small spot of softening near, but not involving the motor structures. The function of these is temporarily impaired, as we know commonly happens in such a case. He has undetected mitral obstruction, a short presystolic murmur having escaped recognition, also not beyond the limits of experience. His medical attendant, who knows that the patient has had constitutional syphilis, believes that the hemiplegia is syphilitic, and at once gives him iodide of potassium. In two or three days there is a distinct return of power in the limbs; the improvement goes on, and in a month the patient is practically well. What will be the strength of the conviction on the part of the practitioner that the hemiplegia was due to syphilis? Will it not be almost impregnable? And yet it would be wholly wrong. Note, moreover, that had the same lesion of the brain been due to syphilis, his reasoning from the effect of the treatment would have been equally wrong, and yet his conclusion would have been right. A great deal of error in medical science (and not in medical science alone) is due to the fact that correctness of a conclusion is transferred to the reasoning by which it has been reached.

The frequence of premonitory symptoms, and of a more or less deliberate onset, without loss of consciousness, furnish diagnostic indications that are often of great importance. They are especially important when some other cause of arterial occlusion co-exists, and especially when a source of embolism can be found. Sometimes heart disease is left by rheumatic fever; the individual contracts syphilis, and afterward has an attack of sudden hemiplegia. We are then unable to avail ourselves of the causal indication, unless we find evidence that one influence or the other has recently been active in the system, as by recent embolism or recent syphilitic processes elsewhere. A deliberate onset is strongly in favor of thrombosis, and therefore of syphilitic disease, rather than of embolism. Most important of all, however, are the premonitory symptoms, and especially headache. This, if severe, shows that a morbid process within the skull preceded the onset, whereas in embolism the intracranial condition is normal until the obstructing particle is suddenly transferred from the heart to the cerebral artery. The practical value of this indication is increased by the fact that it is not only a common, but a prominent symptom. At the same time its absence is far less significant than its presence. This is true, remember, of all symptoms commonly regarded as "characteristic." Their absence proves little when their presence proves much.

Apart from embolism and injury, sudden hemiplegia, coming on between twenty-five and forty-five, is, I believe, very seldom due to any other cause than syphilis, and its occurrence may justly raise a strong suspicion of this disease, even in a case in which there is no positive evidence. But this suspicion is stronger in the case of men than in women, because in women, less rarely than in men, sudden hemiplegia, clearly due to an organic lesion, sometimes comes on when we can trace no source of embolism, and can exclude syphilis. Now and then the onset is accompanied by that form of pyrexia which suggests septicæmia, and actual proof of this may be furnished by the characteristic retinal hemorrhages. A softened clot, in some situation, is the most probable cause of the symptoms, and the hemiplegia may be regarded as embolic. But there are other cases, in which the onset is unaccompanied by any symptom to suggest embolism. I have seen several such cases in girls between eighteen and thirty, in whom, from history, character, and life, inherited or acquired syphilis could be absolutely excluded. The hypothesis of thrombosis in a healthy artery, due to some general condition, seems perhaps the least improbable explanation of such cases. But if the subject of such a palsy happens to be a girl of the lower classes the diagnosis of probable syphilis is not unlikely to be made, and the more readily because we cannot exclude the risk of infection in the case of unmarried women as we can in unmarried men.

With the advent of the degenerative period of life comes an enormous accession of difficulty in diagnosis. After the age of forty-five or fifty is passed, we can never feel sure that we are not in the presence of the changes in the arteries which cause cerebral hemorrhage or cerebral softening. The difficulty is further increased by the fact that syphilitic disease becomes not only relatively but absolutely less frequent, so that the likelihood of it does not help the observer, as at an earlier period of life. Hence, when it does occur it is more likely to be unsuspected and unrecognized. This is indeed one of the chief practical risks of error, for the diagnosis can still be made in most cases, although less surely. The diagnostic indications that are often distinct and important in the earlier period of life are equally common and equally significant in the later period. The significance of some of them is indeed not so strong as in early adult life, but their relative importance is greater, since on them, and especially on those that are still unequivocal, the diagnosis chiefly depends. Their significance is strengthened by a history of syphilis, but this history alone has far less value than at an earlier age. Further, as we cannot exclude syphilis merely because there is no history of it, so we cannot exclude the degenerative lesions merely because degeneration is not manifest. Hence arises the extreme importance of the diagnostic indications afforded by the mode of onset and premonitory symptoms, taken, of course, in combination with the general diagnostic indications that apply to all sudden lesions in late life. The absence of coma, the frequent occurrence of prodromata, or a deliberate development of the palsy, together with the state of the heart, may render hemorrhage unlikely, and the chief difficulty is the distinction from atheromatous softening. In this, as in syphilitic disease, a morbid state of the arteries exists before the sudden onset, and these may cause slight premonitory symptoms analogous to those we have been considering. But headache is not very common, and is never so considerable as it very often is in syphilitic disease. Atheromatous softening, moreover, is essentially a senile malady. Apart from kidney disease or an early degenerative tendency, usually conspicuous, it is not often met with under sixty, and it becomes more frequent as life advances and the syphilitic lesion becomes more and more rare, till at last it is only to be thought of when special circumstances suggest its possibility.

The application of these principles, varying as it does in detail, I must leave; but I may mention to you a case which shows very well their use and value. By a curious and convenient coincidence, this case, one of the best unverified illustrations of this particular diagnostic difficulty that I have ever met with, came under my notice as I was finishing this part of my lecture. Dr. Barnes, of Ewell, asked me to see with him a woman, forty-two years of age, in whom, two days before, right hemiplegia had come on in the course of a few hours, without loss of consciousness. The face was not affected on the right side, and this fact, with distinct weakness of the left side of the face, diplopia, tingling of the left hand, and frequent hiccough, made it almost certain that the lesion was in the pons. The urine had for some years contained albumin and granular casts. In the left eye were the remains of syphilitic choroiditis, and one child had been born dead without traceable cause. For several months she had suffered much from headache, of increasing severity, and had become much stouter.

Here, then, we had evidence of a lesion which was probably softening and not hemorrhage, on account of its position, and of the deliberate onset, without loss of consciousness; hemorrhage into the pons causing more severe initial symptoms, which are, moreover, seldom survived for more than a few hours. Hence, the most probable lesion was disease of the basilar artery, with thrombosis in some of its branches. For such arterial disease we had two adequate causes—renal mischief sufficient to render atheroma probable at her age, and syphilis. There was no evidence of recent active syphilis, and the balance of probability was strongly in favor of atheroma, by reason of the greater average frequency of this affection under these circumstances and causal conditions. But there was one symptom of a positive character, the significance of which no mere probability or improbability could be allowed to influence—the severe and persistent headache. Kidney disease will cause headache, but not such headache as this patient had suffered from, which had kept her awake at night repeatedly, and was often more severe on the left side of the head than on the right. It was a symptom that suggested more than mere atheroma, and distinctly turned the balance of evidence in favor of syphilitic disease. The case was clearly one of urgent danger, as every case is in which there is an active morbid process in the pons, and I confess I had little hope that any treatment would be of avail. But we decided to treat the patient, for a time, on the assumption that the disease was syphilitic, and to increase the iodide, which she was already taking in small doses, to forty-five grains a day, adding a little digitalis to steady the circulation. If no improvement could be observed at the end of a few days, the dose of iodide should be reduced again, lest, if no good could be done to the wall of the vessel, the drug might do harm by increasing the tendency of the blood to clot. But on the third day the patient was distinctly better; the improvement steadily went on, and I hear now from Dr. Barnes that she has fair power in the arm and leg, is free from headache, and has no symptoms to suggest anxiety. Of course, it is possible that the diagnosis was wrong, but its method may not be uninstructive, and the result appears to afford some confirmation both to the principles and to their application. At least we may see in this case how essential it is to arrive at a precise diagnosis of the character of the lesion before the causal indications are considered, and that this need, always present, is especially obvious in those whose age or state of health involves a liability to several lesions. A hasty diagnosis of cerebral hemorrhage in this patient would probably have prevented any significance being attached to the indications of previous syphilis.

We may see also how necessary it is to give its due weight to each indication, how the significance of one may be strengthened by the concurrence of others, and how paramount is the necessity of limiting the influence of mere probability, and not permitting this to counterbalance any distinct and positive indication. Although, as I said, recovery does not prove that a diagnosis is correct, this reservation must be made whenever a patient happily survives. The reservation applies only, moreover, to solitary cases. The test of the reasoning, afforded by the result, may not be valid in the individual case, but in a number of cases combined it will not mislead.

Another point in the differential diagnosis of vascular disease should, however, be specially mentioned, because its peculiar and unusual difficulty occasionally leads astray even those who have had much experience. It is the distinction of the slighter forms of hemiplegia, or other local symptom, from the unilateral symptoms that are common in general paralysis of the insane, even in the early stage of the disorder. Whatever be their immediate cause, they do not seem to depend on an organic lesion, but the symptoms of an organic lesion are sometimes simulated by them in a puzzling manner. The difficulty is increased, and, indeed, is largely due, to the fact that so many of the subjects of general paralysis have had syphilis.

Cases are not rare in which this formidable disease comes on while syphilis is manifestly active; it may even supervene on some actual syphilitic lesion of the brain. Thus the effects of syphilitic vascular disease are occasionally followed by degenerative symptoms of various kind and degree. After recovery, complete or incomplete, fresh symptoms may come on, such as failure of memory or other mental change, or failure in articulation. These symptoms may also commence during recovery from the hemiplegia. Moreover, an attack of hemiplegia, which seems at the time to be due to the common vascular disease, is occasionally followed, in the course of a few months, by symptoms of general paralysis of the insane quite typical in character and course. For instance, a man about eight years after primary syphilis had an attack of sudden hemiplegia, severe, and preceded by headache. In the course of a month he recovered almost completely, and those who knew him observed very little difference from his state before the attack. A little later, however, a slight change in articulation attracted their notice, coupled with some scarcely definable alteration in his mental characteristics. A few weeks later optimistic delusions suddenly burst out; he was about to make his own fortune and the fortune of his friends; the thousands and hundreds of thousands poured from his lips in the flow of magniloquence we know so well, a flow that was soon rendered almost unintelligible by the usual characteristic failure of articulation, and typical general paralysis ran a rapid course to its common termination. In such a case, if the initial hemiplegia is slight and transient, it may be difficult to say whether the attack is due to an organic lesion, or is one of those inorganic attacks that occur in the course of this disease. Indeed, the diagnosis may be impossible except by waiting and watching the case.

Another difficulty is due to the fact that, in general paralysis, attacks occur in which, instead of the sudden weakness of hemiplegic character, there are unilateral symptoms characteristic of a local "discharge" in the brain. When such discharge is so great and so situated as to cause actual convulsion, there is less risk of this particular error, but sometimes there are other sudden symptoms, more equivocal in character, such as tingling, beginning in some part and spreading through the side, and accompanied, it may be, by inhibitory weakness or temporary aphasia. Such attacks have often been ascribed to syphilitic vascular disease. The latter, however, does not cause such definite limited "discharges" as a first symptom. These always suggest a chronic process and not a sudden lesion, and, therefore, in the absence of signs of tumor or of meningitis, the degenerative malady should always be thought of. This is true also if there have been similar attacks, or any change in mind or speech, however slight.

Among the innumerable other diagnostic problems that invite, and almost demand consideration, I can notice only one group, those that relate to the palsies of the ocular muscles. These are among the most frequent manifestations of the influence of syphilis on the nervous system. The long course of the ocular nerves exposes them to many kinds of specific damage. Arterial disease seldom interferes with their functions, and when it does so, it is only by causing a sudden vascular lesion, separable from all others by the complex associated derangement produced, and by the actually sudden onset. Gummata and local meningitis may damage the nerves at the base of the brain, and the locality of the lesion is then indicated by the grouping of the nerves affected; the two processes are sometimes distinguishable by the greater interference with the central organs, or with the motor tract in the crus and pons, and also by the greater frequency of optic neuritis in gummata than in meningitis of the posterior part of the base. Optic neuritis may be readily caused by meningitis when this is in the front part of the base, so that this distinction must depend on the evidence of the position of the disease. Gummata on the nerve trunks themselves can scarcely be distinguished from inflammation. In any part of their course they may be the seat of a syphilitic neuritis, and this is probably the most common cause of an isolated affection of a single nerve or of a single branch.

Thus each diagnosis of this kind involves the careful consideration of the symptoms and their combinations, as well as of their mode of onset and associations. The subject is too large for me to do more than indicate the lines on which it must be approached. It needs, indeed, to receive a thorough investigation in the light of pathological facts that are scattered through medical literature, and which need to be greatly increased in number before our diagnosis can be exact. The greatest difficulties in these cases arise from the fact that ocular palsies are caused by some of the effects of syphilis that are not specific in their nature. For instance, an isolated palsy of one third nerve, in a young man known to have suffered from the constitutional disease, was naturally supposed to be truly specific. But treatment, although prompt and thorough, had no influence on it. One day he suddenly became unconscious, and died in a few hours. We can scarcely doubt that an aneurism, due to syphilis, caused the palsy and his death. But the most frequent non-specific causes of these symptoms are the degenerations. These may set in before the period of true specific lesions has passed, and the ocular nerves constitute one of the regions in which the symptoms of the two kinds most often coincide. More frequent than actual coincidence is the doubt produced by it. Is this palsy due to a true specific lesion, or is it part of nerve degeneration that is a nonspecific sequel to syphilis? The question is of great importance in regard both to prognosis and to treatment. In cases of locomotor ataxy that occur comparatively early, this difficulty is especially frequent. The distribution of the symptoms may furnish an answer, as we shall see in a moment. Sometimes the suddenness of the onset raises a strong presumption that the palsy is tabetic and not specific. The same conclusion is indicated when symptoms quickly lessen without treatment and soon afterward return.

The nature of many of these tabetic palsies is doubtful, but some are certainly due to nuclear degeneration. These, as we saw in the first lecture, are often sequelæ of syphilis. Their distinction from a true specific lesion is another problem of much difficulty—a difficulty further increased by the fact that nuclear affection seems sometimes to follow a peripheral palsy so directly that the symptoms of one pass into those of the other.

The most important characteristics of the nuclear palsy are the association of muscles according to function, its bilateral distribution, and its persistence in spite of treatment. In most true specific lesions prompt treatment has an influence which is absent in the degenerative maladies. I need not trouble you with instances of these features. They are important distinctions both in the isolated degeneration and in that associated with tabes. I may add to them the therapeutic test, which is often of great use in diagnosis, but is of negative significance only when the proper treatment has been employed early.

I have said that the nuclear degeneration sometimes seems to follow a specific palsy, but it also occurs, as we have seen, when there has been no paralysis, so that we are perhaps not justified in connecting the two.

A tendency is often to be observed in the present day to underrate diagnosis, or at least the elaborate diagnosis of which the diseases of the nervous system furnish so many examples. In the face of the urgent needs of suffering humanity, with its mute appeal or uttered cry for the relief we cannot always give, our precise distinctions and elaborate processes may seem like an ingenious device for interesting us while the patient suffers. That such an impression is wholly wrong I need not say to those who hear me now. But the tendency is real, and is reflected beyond our own ranks; and I may give one warning, a warning to myself as well as to others, that we should be always on our guard lest we do anything, by word or manner, that may excite or foster the feeling to which I refer. Diagnosis must come before treatment, and this should make us careful lest we produce an impression that we regard the order of the two in time as also that of their importance—an impression easily produced when the treatment is plain and its methods familiar, while the diagnosis is complex and its processes strange. But surely the diseases we are now considering supply a strong reproof to those who consider that, "to say the least, diagnosis is somewhat overdone." A diagnosis that can be described as "rough and ready" may serve to make our treatment effective in the majority of cases, but there are many in which only a most careful and even elaborate diagnosis will enable us to do all that can be done, and there are not a few in which the utmost degree of care and elaboration is needed to save us from being wrong in our treatment —often wholly wrong, and sometimes disastrously wrong. I could give many instances of this, striking and sad, but it is needless to do so, and it may be unwise. My own conviction is that diagnosis in itself cannot be overdone. To permit diagnosis to be studied and employed, at the expense of treatment, is a proceeding for which no words of condemnation are too strong. To cultivate, at the expense of its application, that which has no other end, should, in medicine, be a sin beyond the reach of pardon. But, as a guide to treatment and prognosis, we cannot well know too minutely the seat of the maladies we are treating, and we assuredly cannot know too certainly their nature. Our art may be all-powerful to save: but without guidance it is useless—a hand to use without an eye to see. When the disease is beyond our present means the same guidance may still be needed to save the patient from that which, without power of good, may not be without capacity of harm, and I believe that there is no point in diagnosis, however elaborate it may be, or however superfluous it may seem, which is not or will not be, in some case or at some time, of definite use and perhaps essential service in enabling life to be saved or health restored.

LECTURE III.

Mr. President and Gentlemen,—To-night we pass to the final ends for which we gain our knowledge—the ability to foretell and to control the future course of the malady. The attempt to define our knowledge, which I proposed at the outset as a chief object of these lectures, will here have results not, I fear, wholly satisfactory. The progress of science involves loss as well as gain. Where the ground seemed firm beneath our feet, and the path before us appeared smooth, we may suddenly find our footsteps insecure, and that the smoothness was apparent only—an illusion due to a too sweeping generalization. When we come to know thoroughly this part of the region of medical science we are now traversing, we find the ground is broken by exceptions, which we cannot ignore except at the risk of certain error and equally certain discredit. It is not long since the diagnosis of syphilis as a cause, and the prospect of adequate treatment, were held to warrant a prognosis absolutely favorable in almost every form of disease so produced. Indeed, I am not sure that it is correct to allude to this opinion only in the past tense, if we may judge from the measures some hopeless sufferers are induced to undergo. We see them subjected to course after course of similar treatment, each to supplement some supposed defect in the previous courses; we see methods employed from which it would seem to be supposed that a drug, administered thoroughly in one way, is still capable of doing more if only it is introduced into the system by some other channel; and we see advice given which suggests a belief that treatment powerless in England will be effective abroad. If we are so bold as to recognize the nature of the opinions that must underlie such measures and such advice, we can scarcely believe that the old view does not still dominate many minds, and that, although it may not often make its appearance on the surface as a definite assertion, it must still exist as an undercurrent of thought, adequate to guide the practitioner's advice, although perhaps not always consciously admitted. Yet the facts that are within the reach of simple observation suggest a very different conclusion. The scars upon the skin or brow tell us more than the mere fact of syphilis: they tell us unmistakably of the imperfect character of tissue-restoration, and that, too often, when the proper elements of a part have perished, their place is and can be supplied only by cicatricial tissue. The internal organs teach us the same lesson. The fibrous nodule in the liver means a loss of liver-tissue perhaps many times the area of the scar, and so with other organs. When such a result occurs in the central nervous system, how can perfect restoration of function be expected or how can it occur? In the liver the effect on loss of function may be inappreciable, but in the nervous system each cell and fibre has its special use, and the loss of a few may have effects on function not only appreciable but conspicuous. In truth, in a very large number of the cases of such lesions, complete recovery of function does not and cannot take place.

Yet also in a large number of cases the symptoms pass away. Between entire recovery on the one hand, and entire persistence of the symptoms on the other, we have every degree of partial recovery. The course of the disease in any given case may be anywhere between these two extremes, and if we are to foretell it aright we must recognize this fact, and we must learn where to seek and how to find the indications that will guide us. Every prognosis, therefore, must be a matter of special and individual consideration, just as diagnosis must be. No rule can be laid down applicable to all cases, and the rules that can be made for any groups of cases are limited and partial. In each group, indeed, the general prognosis varies, but in each group also the individual case must be dealt with as a separate problem. So stated, the task of acquiring any adequate prognostic power may seem both difficult and long; but, happily, the power is easily acquired, and its use is simple. I said that "as much individual consideration is needed as in the process of diagnosis," but I should have been nearer the truth had I said that the same individual consideration will almost suffice. Did the statements regarding the process of diagnosis I made in the last lecture appear to you exaggerated? Surely they will seem within the truth when we realize that a full and correct diagnosis is not only essential for, but actually almost involves, the most complete and correct prognosis it is possible for us, under the circumstances, to give. Realize the nature of the lesion; realize that the symptoms depend not on the specific process, but on the simple effects that are thereby produced in the nerve-tissue; picture to yourself the changes that have taken place, consider how long they have existed, and you have only to ask yourselves how far it is likely that the mechanism of damage can be removed, and how far the nerve elements can recover from the damage they have sustained.

Let me repeat the injunction, "picture to yourselves the changes that have taken place," for its importance is profound. Acquire the habit of forming a mental picture of the morbid process and the way in which the lesion is doing its work. You will find it of unspeakable service to you in giving you a firm grasp of the nature of pathological process, a grasp that will help you to a prognosis as adequate and a treatment as wise and full as the nature of the case permits. This is true not only of this, but of all diseases. To pass to the special prognosis of these diseases, only that of the true specific lesions can be now considered. It is only in these that the fact that disease is due to syphilis has an extensive and considerable influence. The lesions that are doubtfully specific, and those that are not specific in their characters, do not often respond to the special drugs, and their forecast is not influenced by their causation. The specific processes alone will more than suffice to occupy us. But here, again, we meet at every turn the importance of precision in diagnosis. We must decide the probable nature of any lesion, decide whether it is specific or non-specific, before the questions of prognosis can be even raised; and we must further decide, as far as we can, the precise form of lesion that exists, before any attempt at prognosis is possible. This done, the prognosis becomes at once comparatively simple. The fact that a separate answer must be given in each case does not involve, therefore, a mental process essentially different from that of diagnosis. The special prognostic element, the process by which we read the future in the present and the past, is much the same in all lesions. In brief, we have a varying diagnostic process, but a prognostic process that only differs in so far as its color and outline, and not its real form, vary with the diagnosis. This, as, indeed, most of the statements I have made in these lectures, are merely familiar facts in perhaps a fresh form. It is strange how far certain principles, when they are definitely grasped, will carry us, and how much they will effect in their various applications. Not seldom a change in the form in which a truth is presented makes it almost equivalent to a new truth. Hence I venture to be so persistent in my reiterations, for repeat I must, again and again.

Prominent above all the others in prognosis, therefore, we must remember the two leading principles, by this time, I hope, familiar to you: first, the dependence of symptoms on the simple processes in the nerve tissue, and not on the specific element; secondly, the fact that it is the specific process only on which our treatment exerts a direct influence. So the first conclusion that we must seize and hold fast is that treatment can never influence the symptoms directly. The simple processes that cause the symptoms are scarcely at all amenable to direct treatment, and not at all to the specific treatment. The utmost we can do is to try and remove the specific process, and so allow the nerve elements to recover, if they can. Here, then, we see the essential principles that must guide every prognosis. Is the specific process such, and in such a stage, that we cannot remove it wholly or partially? Is it to be expected that such removal of it as we can effect will be followed by such recovery of the damaged nerve tissues as will permit a return of function? This is the method by which, and by which alone, in every form and degree of lesion, our prognosis must be obtained. The process, thus, is simple. Its application is not difficult, but it needs the same exactness, the same attention to every step, that we have seen to be requisite in every practical proceeding in connection with these diseases. No short road can be found here, any more than in diagnosis. We may make each step sure without any extensive special experience, but we can only do so by patient care. Any senior student who has been properly trained may, in the majority of the cases of syphilitic disease of the nervous system, give a prognosis that shall be nearly as correct as could be given by a physician of large and special experience, and each must employ the same process.

The first question then is, how far can the specific process be influenced by treatment? We caught a glimpse in the first lecture of the variations that exist in the susceptibility of the different processes to treatment. We saw that hyperplasia is the great characteristic of visible specificity, and that it is this element in the process over which treatment has most influence. Inflammation, even when it is due solely to syphilis, seems to vary much in its susceptibility. Here, however, our vision is limited by the fact that we cannot discern, during life, how much of this apparent inutility of treatment is due to the lack of influence over the specific element, and how far it is due to the fact that the nerve elements *cannot* recover. In chronic inflammation it sometimes seems as if we cannot arrest the true specific process, but in acute and subacute inflammations the facts we can perceive and the analogy of similar processes elsewhere make it probable that the second explanation is the true one; that it is the persistence of the damage to the nerve elements, and not of the lesion producing the damage, which causes the persistence of the symptoms.

The first important consideration, however, is certainly the fact that the tissue-formation of syphilis is the element that can be influenced with most frequency, with most certainty, and in greatest degree. But it is only in the early period of the process, when the tissue is still soft, that it can actually be made to disappear. When it caseates it may also disappear, although the removal of the degenerated particles is perhaps not due directly to our treatment. When and where the process of fibroid transformation has set in, it is doubtful whether our treatment does more than aid this process, which is really cicatricial in nature. The tissue that is produced lessens in amount whether we treat the patient or not, but apparently only as the shrinkage takes place which all scar tissue undergoes, a shrinkage by which the elements occupy less and less space as they change from the original cells, or their original form, into inert fibres.

In the process of sub-acute and chronic inflammation, moreover, the effect of treatment through the specific element is largely dependent upon the amount of new tissue formed, and the stage of development which this has reached. Inflammation, however, causes also direct damage to the nerve elements, proportioned to the acuteness of the process. In acute inflammation it is great, and seems to constitute the most effective element of this process. We saw that in such acute inflammation there are commonly no true specific features. Our treatment has only to bring the acute inflammatory process to an end, and if this can be effected speedily enough there is nothing further for the special treatment to do: there is no new tissue for it to remove. When we have arrested the process of inflammation all we can do in our treatment is to wait while the nerve structures recover, and all that we can do in our prognosis is to endeavor to form an opinion as to whether they can recover or not. But when the inflammation is chronic, or, if acute in onset, has lasted for a long time, we have to consider also the tissue that may have been formed, the extent to which its removal is possible, and the effects which it has produced on the nerve elements, as well as the degree to which they can recover.

These, then, are the chief elements that have to be taken into consideration in forming a prognosis. But, in addition, there is a question, subordinate to the others in actual position, but at least equal to them in importance. This question is the state of the new tissue in connection with its relation to the nerve elements. You will see what I mean in a moment. As I have just said, when the fibroid transformation has set in, treatment does not seem capable of removing the new tissue, or doing more than hastening this cicatricial process. But this process involves, as we have seen, contraction of the fibres. If the new tissue merely compresses the nervous structures from one side this contraction is as effective as absorption in relieving these structures from the damaging pressure. But whenever the new tissue encircles the nerve elements and encloses them, the compression is actually increased by it. Thus the healing of the syphilitic process, the recovery of the essential syphilitic lesion, may mean that the damage to the nerve structures is maintained, and so also are the symptoms produced by that damage. After a time, it is true, there is often a further improvement in the symptoms. When the process of contraction is at an end, perhaps even before it comes to an end, nerve fibres, at any rate, seem to be able, in some degree, to accommodate themselves to the new conditions, and although greatly narrowed, may yet regain some capacity of conduction, which is sometimes considerable in amount and goes on increasing for a long time.

We see a very pertinent illustration of these important facts in the eye, in the case of common optic neuritis. It is common for sight to fail more during the process of subsidence of the neuritis than when the inflammation is at its height. The inflammation produces a large amount of new tissue in the subsidence of the optic nerve at the papilla. This infiltrates the nerve extending between the fibres, and the contraction of this tissue necessarily involves great compression of the fibres. But when the subsidence has become considerable, when most of the swelling has passed away, some improvement in sight commences, and often it goes on slowly for many months, and ultimately reaches a very considerable degree. But it is also common, when the compression of the optic fibres is great, for the ultimate recovery to be incomplete, and occasionally the amount of vision that is regained is even at last very small. In many of the syphilitic lesions of the nervous system the processes which we can see so clearly in the eye, and estimate so accurately by our tests of vision, certainly go on in like manner, and with corresponding effects upon the function of the parts.

Hence it follows that whenever there is reason to suppose that there is new tissue which has begun to undergo, or which has already undergone, the process of fibroid transformation, and that this new tissue is so placed as to surround either the spinal cord, the nerve roots, or nerve fibres, our prognosis must be influenced by the considerations I have just mentioned. In all inflammations that are not actually acute or are not brief there is this new tissue. If the inflammation has lasted for a few weeks before treatment has been effective we cannot assume that the new tissue will be removed. The amount of new tissue will be proportionate to the intensity, extent, and duration of the inflammation, of which other indications enable us to judge, and we must consider how far the position of the tissue is likely to entail this encircling compression. Hence, until we can see what influence treatment has upon the symptoms, our prognosis must be deferred, at least so far as details are concerned. If improvement is absent or slight, or when improvement begins to flag, we must expect that the degree of disturbance of function which then remains will, for a time, lessen only in slight degree or at a slow rate; that a little further improvement may take place in the course of time by the functional adaptation of the nerve elements to their new condition; but that some amount of the impairment of function is not unlikely to remain. We must recognize also the important fact that when once the early and distinct improvement that follows our treatment is at an end, or even when it has distinctly begun to slacken, it is not likely that any further improvement the case will show will be in any degree the result of anti-syphilitic treatment. I believe that these general principles are absolutely true, and I am sure that their recognition in practical work is of the utmost importance. Their application to any individual case will furnish an accurate and adequate prognosis, and as I said, this application requires rather care than practice, patience than special knowledge.

Thus three points in every morbid process have to be considered—the nature of the tissue in the specific lesion, the mechanism by which the nerve elements are damaged, and the character of this damage. We must consider the nature of the tissue to determine the effect our treatment is likely to have upon it; we must consider the mechanism by which the nerve elements are damaged to determine how far the changes we are able to produce in the specific tissue can relieve the nerve elements; and we have to consider the nature of the changes in these nerve elements to determine whether it is or is not likely that they can recover function if completely relieved from the morbid influence. We have seen that it is by the removal of new tissue, which is still in the early stage, by reducing the bulk of such tissue when it is exerting compression, and by checking the process of pure inflammation, that our treatment is effective. We have seen that it is when the compression is from one direction that the reduction in bulk of the syphilitic tissue has the greatest effect in releasing the nerve tissue from the influence of the syphilitic process, and that when the new tissue presses on them from all sides the diminution in its bulk is more than compensated by the constricting contraction of the tissue.

A word or two more seems needed regarding the third element, the changes in the nerve elements themselves. In pure acute inflammation, even when due to syphilis, this is the only point to be attended to. In the subacute and chronic inflammation it is secondary to the considerations regarding the new tissue which have just been mentioned.

Note, once more, that these changes are simple and not specific, and that it is on them that the symptoms depend. They may be the simple breaking up of the tissue elements, seen in simple inflammation and also in necrotic softening of the brain from syphilitic disease of the arteries. The prognosis of this we considered so fully last time that I need not speak of it further. Except in this process of necrosis, and in the very rare pure acute inflammation, I do not think you will find any certain instances of change in the nerve elements in specific processes without a production of new tissue. Secondly, we have the effects of compression already mentioned. Simple lateral compression readily arrests function, but it seems to do so by producing alterations in the nerve elements that can be recovered from to a larger extent than is possible in any other form of damage. Whether the axis-cylinders persist or not, the fibres seem to be able to regenerate even after they have been exposed to such pressure for months. Lastly, in the damage produced by so-called "hyperplastic inflammation" there seems to be both compression by new tissue and also the alterations that result from the inflammation itself, which we have already considered. The contraction of tissue that encircles the elements keeps up the compression originally caused by its bulk; hence this, and also the persistence of the inflammatory changes in the nerve tissue, cause the effects of such hyperplastic inflammation to persist long after the activity of the inflammatory process itself is at an end, and even long after the treatment has ceased to exert any influence over the morbid process.

Except for another point, I might content myself with this outline of the general principles of prognosis. Time fails me to illustrate its use, but I may briefly enumerate the general character of the effect which these considerations have upon the special prognosis of the chief kinds of lesion. Afterward I will mention the one remaining point that has to be considered—the modification of prognosis entailed by the *irritation* which some lesions cause.

The practical influence of these facts upon the chief lesions is as follows: syphilitic growths—gummata—can be removed more completely than any other lesion, and when their influence is exerted by simple lateral pressure the resulting damage to the nerve structure is recovered from in the greatest degree. Hence, in such cases, the prognosis is good, unless the duration of the damage is very long, and its degree is very great. When the growth infiltrates, however, the problem becomes similar to those met with in some inflammations.

In all forms of meningitis the prognosis is good except in so far as relates to the damage to nerves and nerve-roots which are surrounded by the newly-formed tissue, and also the damage to the spinal cord when the dura mater is greatly thickened. In all these cases the effects and the symptoms commonly lessen up to the point when the contraction of the tissue comes into operation. The effects of the inflammatory damage, and of the early compression, lessen, but a residuum remains in most cases, and is increased in some of them. Although further improvement is not uncommon after a time, some permanent effects may remain. When there is reason to suspect inflammation of the cranial nerves—syphilitic neuritis—the prognosis is chiefly influenced by the duration of the morbid process. If the inflammatory products are in such a stage that they can be removed the prognosis is good; they will recover conducting power. If, however, there is reason to suspect cicatricial transformation of the new formation into fibroid tissue it is improbable that the degree of improvement will be sufficient to permit *severe* symptoms to pass away entirely. The same general prognosis is true also of infiltrating growths. In true acute inflammation that is treated early and arrested early, the prognosis is the same as in similar inflammations that are not specific. The recovery or persistence of the symptoms depends on the recovery or persistence of the changes in the nerve elements which are produced by an intense inflammation. Such changes are simple, and not specific.

The point which I have left to the last is the prognosis of the symptoms of *irritation*, which forms so large a part of the manifestations of many lesions. The irritation may be that of nerve fibres, causing pain, or that of motor structures in the cortex of the brain, causing convulsions. The irritation of nerve fibres is due to the influence of inflammation in the first instance, and secondly to the influence of the residual compression I have spoken of, and it has a course analogous to the other effects of this compression. It therefore need not detain us. But the irritation of the cortex which is manifested by convulsions is of great import in relation to the problem of prognosis, and needs special consideration in every case in which its action can be traced. Convulsions are often produced by a syphilitic lesion on the surface of the brain; they may persist after treatment has done all that is possible in removing that lesion. Such convulsions are usually ascribed to a cicatrix on the surface of the brain. They are commonly of local commencement or distribution, and a cicatrix does commonly exist at the spot to which the local commencement points as that in which the " discharge " commences. But general convulsions—always general in commencement—may be the sequel to what seemed, from the symptoms, to be diffuse general meningitis. The theory which is commonly held, that a cicatrix causes such symptoms by persistent irritation, may be correct, but it is not easy to understand perpetual irritation by a stationary cicatricial process, unless there are such adhesions to the bone as involves repeated and perpetual mechanical disturbance. The facts are equally explicable on the assumption, which seems to me much more probable, that the " discharges " originate from slightly damaged nerve cells, and that the discharging tendency is perpetuated by the residual influence of the discharges themselves. This view is quite consistent with the effect of the excision of a cicatrix in arresting the attacks, since the removal of the scar entails that of the nearest and most unstable cells. Be the cause what it may, the effect is certain and important, though the symptoms may persist long after all other symptoms of a cortical lesion, and may even constitute in effect, although not in origin, a variety of epilepsy. These persistent fits, especially when general, are not influenced by anti-syphilitic treatment. Hence their occurrence must always cause a special concern, and should tinge your prognosis with a special caution.

A general survey of the treatment of syphilis, as well as the observation of individual cases, can scarcely fail to impress one important conclusion on the impartial observer. It is difficult to express this conclusion in language not open to misconstruction. Stated in its baldest form, the conclusion is that syphilis is an incurable disease.* Stated in the form least liable to misinterpretation, and with the limits of exactness, it is this: There is no real evidence that the disease ever is or ever has been cured, the word "disease" being here used to designate that which causes the various manifestations of the malady. The short statement that "syphilis is an incurable disease," is legitimate if we recognize that "incurable" means that there is no proof of cure. I do not put this opinion forward as in any way novel; indeed, I feel that on this and on many other points in these lectures I express opinions that are held by many others, some of which have been well stated by others,

---

* This statement has been widely objected to. It should be noted, however, that it is perfectly consistent with the recognition of the fact that the lesions of syphilis have been and are readily removed by treatment.

and all of which, I cannot doubt, have been expressed more than once.* In current works on syphilis the fact of incurability is clearly admitted, although even in these it is not quite compatible with the language here and there employed. But the fact is certainly not accepted by the profession at large in the manner that seems to me to be indicated by the facts. It is, I think, most important, for practical reasons, that it should be understood by all that, as far as evidence at present goes—whatever power we may hope and anticipate that the future will bring—no treatment, however thorough, will bring syphilis, as a disease, to an end, so that the patient does not suffer again from any of its direct effects. In this sense (and it is the only proper sense of the words) the assertion seems to me to be open to no question. Yet the fact is very remarkable. It is strange that this should be true of a disease the effects of which are so largely under control. We can, under favorable circumstances, remove most true specific lesions, we can bring to an end most true specific processes, but the arrest or removal of the manifestations of a disease is a very different thing from its cure.

We can, indeed, do a little more than remove the manifestations and effects of the disease; we can restrain

---

* I regret that it has not been possible for me to ascertain the extent to which these and other opinions have been before expressed. Hence I feel that the only course at once just to others and safe for myself is to ask you to regard nothing in these lectures as having any claim to originality. As I said at the outset, my desire is to render more definite the knowledge that already exists, and not to bring new knowledge into existence.

the activity of the virus, whatever it be, that causes the manifestation and effect. But when we speak of the *cure* of a disease we mean that its essential element, the virus that lies behind all its symptoms and consequences, that which is the persistent cause beneath the transient effects—we mean that *this* is made to cease, is ended once and for all as a morbific agent, so that it never again disturbs the system. In this sense I believe that it is literally correct to say that we have no evidence that syphilis ever is, or ever has been, cured.

I can scarcely doubt that so absolute a statement will seem, even to some here to-night, unjustified and even mistaken. Some of you may call to mind cases that seem to contradict it, cases in which a patient was suffering from symptoms of the disease, and is known to have been afterward free from any indication of the malady. Such cases do occur; they may be numerous. Far more frequent, however, in the experience of most observers, are cases in which the immunity that follows a course of treatment is not permanent, in which other symptoms reappear at a later period. In very few of the cases regarded as "cured" has the patient been kept under observation long enough to justify the positive conclusion that the disease was at an end. Nowhere has this fact been recognized more clearly than in the admirable work of Hill and Cooper. Cases are common enough in which the hope of cure produced by several years of perfect freedom was ultimately disappointed. Of the patients with late syphilitic lesions that come under the observation of the physician, many, very many, have undergone thorough treatment for the early symptoms. It is difficult, I may say that to me it is impossible, to compare the history of the various sufferers from syphilitic lesions, to compare the course of the disease in cases in which treatment was early and thorough with those in which no treatment was employed, and not to question whether early treatment has yet been shown to have any appreciable influence in preventing the late symptoms of the disease. But if this be so, how are we to explain the cases in which the malady has seemed to cease after a course of treatment, or those in which it has ceased so long as observation has continued? I have already hinted at the explanation which I believe to be the true one. We saw in the first lecture that if we compare a series of cases, we meet with many in which the primary affection received no treatment at all, and yet the later symptoms were of a trifling character. There is the strongest reason to believe that many other cases occur in which late symptoms are altogether absent. Some facts that have an instructive bearing on this subject come to light in the histories of patients in the medical wards of a hospital or in the out-patient room, suffering often from diseases that have no relation to syphilis. It is not uncommon to meet with those who have unquestionably suffered from primary syphilis and from slight secondary symptoms, but who took little or no notice of them, and who have never been treated, have had no later manifestations. I have mentioned the series supplied to me by Dr. Radcliffe Crocker, and which I hope he will extend and publish, for their importance is very great. In this series there are many cases in which notable late syphilitic skin eruptions occurred without any recognized previous manifestations of the disease, and even many in which the primary lesion had been unnoticed by the patient, and was certainly never treated. As I have said, there must be many more with a similar latency of the early stages, with a similar absence of treatment, who passed, not merely six or ten years without anything to call the malady to notice, but who never observed, through life, any manifestation of the disease to tell them that they were or had been its subject. Take, again, the other class of cases I referred to—the immunity of mothers of syphilitic children. They are certainly syphilitic, and yet many of them have never been treated and never suffer. I believe that the variations in the general course of the disease, in those who have been treated, find parallels in the course of cases that have not been treated. I suspect that if a large number of each class could be compared the discrepancy between the two series would be found to be surprisingly small, and that it would not be difficult to find in each series a corresponding proportion of cases, on the one hand, in which the disease without treatment was brief and trifling, and, on the other hand, in which the disease with treatment was prolonged and severe. All the facts, impartially examined, seem to me to have one meaning, clear and unmistakable to those who choose to read it. The extreme variations in the natural character and course of the disease have not been sufficiently considered and allowed for by those who are under the impression that the cure of syphilis is certainly within our power. This impression is an illusion produced by variations in the course of the disease. In some patients the disease is brief; the symptoms that bring the patient under treatment may be the expiring effort of the virus, and the spontaneous cessation of the disease is ascribed to the treatment which was adopted. In other cases the tendency to further manifestations, at some future time, near or distant, may be so strong that in spite of the most energetic treatment, long continued, many times repeated, wave after wave of activity brings to the unhappy subject year after year of varied suffering, and sometimes his life ends before the disease.

I have put the matter strongly in order to emphasize the essential truth. But certain qualifications are desirable, although they do not touch the fact itself. One qualification is that while we have no proof that we ever can or ever have cured syphilis, it is possible that we do come very near the achievement, and even sometimes attain it. I said that some manifestations are probably—indeed, I might say certainly—the expiring efforts of the malady. It may be that in some others the symptoms that bring the patient under treatment are only the penultimate manifestations, and that some further consequences, which would occur were it not for the treatment, may be averted. Then we do arrest the symptoms altogether. In such a case the malady may be said to be cured; but the cure is at best limited, and cannot be regarded as more than hypothetical. The second reservation is one rather of words and theory than of fact. There are some who hold that the duration of the disease is much more brief than that of its manifestations, that these later symptoms are indirect consequences, either in their mode of production or in the mechanism by which they are caused. The statement I have made is one that is independent alike of pathology that is known or any theory that may be held. Whatever be the intimate and ultimate pathology of the later symptoms from the point of view of practical medicine, they are part of the disease, and no question as to their origin or their relation to the virus can alter the fact that the disease, as a disease to be dealt with, and not merely to be thought about, is not, in the proper sense of the word, a disease that can be cured. Speculations as to the origin of the lesions, and the essential nature of the malady at different periods, have no influence, it seems to me, on the practical problems to which we must now confine our attention, although it may be that these practical questions have an important bearing on these speculations. Although the course of the disease in its relation to treatment justifies the doubt whether it is ever cured, except in this narrow and partial sense, the question arises whether the removal of the lesions, and the repression of the activity of the essential elements of the disease, have an influence on the course of the affection. It is difficult here to resist the seductive attractions of theory and of analogy. But the question can be answered only by facts, and we ought not to try to obtain the semblance of an answer in any other way.

The conclusion that the essential element in the disease resists treatment, and runs its course uninfluenced by our efforts, is in harmony with what we know of other specific diseases due to a poison introduced from without and communicable from one person to another. There is not any fact whatever to show that a single disease of this kind can be cut short. The course of the acute exanthemata cannot be arrested by any means at our disposal at any stage of their course, and the same seems true of this chronic exanthematous disease. This is eminently true also of the disease that stands perhaps nearer to syphilis than any other known malady—leprosy.

Here we catch another glimpse of the strange duality of the disease. And yet the word "strange" is scarcely accurate. In syphilis we probably only see that which exists in all maladies of like nature, but in most others, to a large extent, escapes our notice. What is it that treatment influences in so remarkable a manner? What is it that remains in spite of the treatment—remains, to assume from time to time a fresh activity, and to cause renewed manifestations? We say, in general terms, that treatment removes the effects of the disease, the local lesions it produces. Besides this, treatment seems to be capable of repressing, and even suppressing, the activity of the *materies morbi*. Take this one fact, evidence of which you will find presented with care in the work of Hill and Cooper: A syphilitic woman who would bear a child certainly diseased, and probably dead, can by treatment be made to bear a child that at the time of birth shows no trace of the disease; yet even such an influence on the morbid process in the mother does not prevent her from again suffering. If we are able to hold securely the outline of a general truth only by means of the point of an hypothesis, we may consider that our drugs destroy the developed and developing organisms, and fail to influence the germs of these. But much remains obscure, when this or any other hypothesis that we can frame has done the utmost that it can to clear our view, and we must wait, perhaps for long, before we can understand the real reason why we are able to do so much, and yet can do no more.

The general treatment of syphilitic lesions does not come within the limits that I have set to these lectures, and even if time permitted me to make an exception on account of the importance of the subject, I should still hesitate to take up your time with that which every student is taught. The few minutes that remain I may best spend in pointing out certain points on which we need to have a far firmer foundation of observed facts than has at present been supplied.

One question on which there is some difference of opinion is regarding the respective power of mercury and of iodide of potassium, which is represented in its extreme forms by the custom of some physicians to give only iodide for the late lesions, and the opinion of others that mercury alone is effective, and that iodide has only the power of bringing into an efficient activity the mercury that may have been deposited and held inert in the various tissues. The only point chiefly at issue, then, is the influence of iodide. I have on this subject to range myself with the majority. It seems to me impossible to see much of the true specific lesions of the nervous system, and not to believe that over most of them iodide has a powerful influence. The symptoms of a gumma, for instance, lessen and pass away under the influence of iodide of potassium as speedily and as completely as would seem, *à priori*, to be possible with any agent. I am inclined to think that when a lesion is distinctly influenced by either drug, if this is given freely, the effect of one is as great as that of the other; and in most circumstances I doubt whether it is necessary to follow the iodide with a course of mercury. But this is only an impression. The physician cannot gain from his cases any really trustworthy *conviction*. We need more detailed observations by those under whose observation the treatment of lesions in the skin and other parts can be *seen;* we need observations, superfluous for them but most important for us, as to the relative effect of different kinds of treatment, and the time needed for the complete removal of specific processes. And we need, in such observations, careful discrimination, as far as is possible, between the actual effect of the drug and the changes which result from that effect, but need time for their full completion. It is very easy for physicians to mistake the remote effect of the treatment, the slow improvement in symptoms that goes on for a considerable time after the syphilitic lesion has been removed, for the slow removal of this lesion.

At the same time the question needs asking, and needs an answer: Are there late syphilitic lesions over which iodide of potassium has no influence, and which yield to mercury? Some years ago this question was impressed upon me by the case of a woman who had a firm mass deeply seated in the posterior triangle of the neck, compressing the lower nerve roots entering the brachial plexus, and causing paralysis of the muscles of the forearm and hand. She presented unmistakable signs of constitutional syphilis, including the loss of almost the whole of the soft palate, which had been destroyed by ulceration several years before. Iodide of potassium was given in large doses for about six weeks, but without the slightest effect on the mass. Perhaps during the following two months it did not get larger, but it certainly did not get less, nor did the pain which was caused by the compression of the nerve roots. Mercury was then given, and as soon as its influence was established the mass began to lessen and the pain ceased, and in a few weeks the neck had resumed a normal state. The lesion was no doubt the chronic syphilitic periadenitis not uncommon in this region. Except for one possible source of fallacy (to be mentioned later), this case seems to be conclusive evidence of the possibility of this difference.

I have not met with any case of intra-cranial disease in which there was reason to believe that mercury was successful while iodide failed. In the case of gummata I have found, as a rule, the effect of iodide perfectly satisfactory, as great as seemed possible, and with entire removal of the symptoms, except in cases in which such complete removal could not be expected. Syphilitic inflammation is, I think, better treated with mercury, since this has far more influence than iodide over the process of inflammation, irrespective of its nature and cause, but I cannot say that this opinion is based on observations free from sources of fallacy.

With regard to the methods of administration of mercury I have nothing to say. The old method of inunction seems to me to bring the patient under the influence of the drug as speedily as it can be done with safety, and with a certainty incomparably greater than the administration by the mouth. I have been deterred from a trial of the hypodermic method because the published evidence seemed to me not to afford any satisfactory proof of superiority, being destitute of the element of comparison essential to such proof, and because this method seems to afford an opportunity for psychical influence not free from risk of that which is undesirable. But I would not for one moment suggest that such an influence has entered into the motives of those who have used this method.

There are two points in the treatment of syphilitic diseases of the nervous system on which I think a word of caution is urgently needed. The first is regarding the prolonged administration of anti-syphilitic drugs, especially of iodide of potassium. By "prolonged" I mean exceeding from six to ten weeks. I believe that full doses in this time will effect all that can be thus achieved in the removal of the syphilitic process. But here, as I have just said, we want facts that are visible—numerous and carefully observed—to guide us in our conclusion. I do not say that the symptoms will have disappeared.

It cannot be too firmly remembered that symptoms are due to changes that are not syphilitic—changes in the nerve elements secondary to the syphilitic disease, but so far independent in course that they may persist long after the specific lesion is at an end. Hence, the fact that the symptoms have not yet ceased is no indication that the specific lesion is not entirely removed. Here, then, I would range myself with the minority—a minority fast growing in size, especially in Germany—with those who hold that the long-continued treatment by small doses of mercury or iodide is a mistake, great and dangerous; who hold that treatment of any true specific lesion should be energetic, but should continue only a little longer than is necessary to remove the lesion, repeated, it may be, after an interval occupied either by tonic treatment or by the other of the two chief drugs. If iodide is continued, as it often is, during many months (and much more, as it sometimes is, during years), there is a danger that the system and tissues of the patient may become so accustomed to its presence that the drug will no longer hold in check the syphilitic processes. Probably we may still influence the lesions by increasing the dose, but this process must have limits, in practice if not in theory. Such prolonged and augmented treatment may do definite injury to the patient's health, and even then may fail to effect the desired object. If the poison is an organized virus we might *à priori* expect this result. By long-continued gradual alteration in the conditions, low organisms, as Dallinger has shown in his remarkable experiments, can be made to endure influences that would at first be fatal to them. By slowly raising the temperature, they will not only live, but will flourish, at a degree of heat which, had it not been for their acclimatization, would kill them in a few minutes. Still, the question is not one in which we can be guided by theory, far less by an analogy possibly remote. There is, however, a positive danger in this method, and it is especially great in the practice to which I referred at the beginning of the lecture, that of repeated courses of energetic treatment to remove residual symptoms that cannot be thus removed, because they do not depend on any residual specific process. The test of experience alone can show the *extent* of this danger, but, at the same time, the *fact* does not need for its proof the evidence of accumulated experience or numerous observations. A single clear instance will suffice to establish it, and no number of negative instances would disprove it. We know how erratic the course of syphilis is, and how great must be the variations in the state and tendency of the virus in the system. In one case, if the energy of the disease is subdued, it is for a long time tranquil, while in another the tendency to fresh development soon reasserts itself. Hence it may readily happen that a series of cases leads to a delusive confidence in the safety of a course of treatment, which is, nevertheless, now and then fraught with danger.

The following instance clearly proves that the danger is not imaginary. Many years ago a man came under my treatment with symptoms of local chronic meningitis about the pons. Ten grains of iodide of potassium were given, and the symptoms rapidly lessened. Slight residual symptoms, however, remained, for which iodide was continued, and he took it regularly for about four months. At the end of that time some obscure cerebral symptoms developed, the syphilitic nature of which was doubted because they had developed in spite of the iodide. He was admitted to the hospital, and the iodide was continued and slightly increased, but the symptoms rapidly developed, spinal symptoms were superadded, and in a few weeks he died. Post-mortem we found the remains of local syphilitic meningitis at the base, as had been suspected, and also a second syphilitic gumma in the cerebral hemisphere, and another in the spinal cord. Here, then, we had clearly a new development of that syphilitic lesion over which iodide has most influence, although the patient was at the time, and had been for long before, continuously taking the drug.

I have seen similar instances, both with iodide and mercury, but this case impressed me so as to preclude the occurrence of fresh illustrations in my own practice, and I do not care to adduce evidence not observed by myself throughout. Especially in early syphilis, however, I have been satisfied that the virus may resume activity during the continuation of mild mercurial treatment. Here there is a source of a possible fallacy in the case I mentioned just now, in which iodide had no influence on the syphilitic process in the neck. It is possible that the patient had been taking iodide for a long time before she came under my care. There is no evidence of it, but the point was not investigated, and must remain doubtful.

I would add only one other remark. If it is true that we cannot cure syphilis, it is most important to consider how it can best be kept in check. This is why the fact of incurability, if true, is so important. A mistaken belief in curability may dangerously hinder attempts at prevention. If no present treatment can prevent future development, then it is wise, whether they come or not, to anticipate them. I think the custom sometimes recommended is good—that every syphilitic subject, for at least five years after the date of his last symptoms, should have a three weeks' course of treatment twice every year, taking for that time twenty or thirty grains of iodide a day. If this practice were adopted generally, is it not reasonable to anticipate that grave lesions would be much more rare?

The idea that residual symptoms are necessarily proof of residual specific lesions, which can be removed by a renewal of treatment that has already been energetic, is not only destitute of proof, but is, as I have said, dangerous. I have alluded already to the ease with which a mistaken conclusion can be reached. The danger depends on the tendency which I have more than once mentioned, and must again advert to in its therapeutical aspect—the tendency to degenerative changes in the nervous system, presented so frequently by those who have had syphilis. The danger is considerable both in those who have had true syphilitic lesions and in those who have not. These degenerations, as we have seen, are not checked by specific treatment, and there is some reason to think that they are not rarely accelerated by such treatment, especially when it exerts a depressing influence on the general health. I have seen many instances of this, especially from energetic mercurial treatment. In most cases of the kind, the evil has been done by the repetition of mercurial treatment, in the hope of removing symptoms which were not, and could not be, removed by such treatment.

There is one other important element in the treatment of luetic lesions in the nervous system that is often forgotten. We have to do with a syphilitic process, and with the damage to the nerve tissues which the process causes. Thus there are three elements in the affection —the specific process, the damage to the nerve tissues caused by that process, and the symptoms that are due to this. It is important to recognize each of these elements, and to consider, in arranging our treatment, how far the two pathological elements need special measures. In arranging the treatment of syphilitic lesions we must never forget, any more than in diagnosis and prognosis, the simple changes in the nerve tissues on which alone the symptoms depend. To confine our attention to the syphilitic element in treatment is to neglect measures that may make an appreciable difference in the ultimate result. The cases are, it is true, rare in which we are able to appreciate the effect of neglect on the one hand, or of recognition on the other. In the improvement that follows the removal of the syphilitic process, we are commonly unable to discern that which may be due to collateral measures. A chief bane of modern therapeutics is, it seems to me, the demand for proof where proof cannot be forthcoming, and the tendency to reject that of which the evidence is not clear, when the absence of such evidence is often no ground for inferring the absence of benefit. In every case, all measures should be employed that are suggested by what we can learn of the nature of the morbid process and of the known tendency of therapeutic agents, in confidence that they cannot be altogether without influence in bringing about the ultimate result, and satisfied if now and then we can distinctly trace their beneficial action.

Here I must end. To me, and if I have been in any measure successful in achieving the result that has been my aim, to you also, the subject cannot but grow in importance the more it is studied; its vast proportions seem to loom still vaster. through the mists that obscure its features and prevent clear vision of its outline—still to us unlimited. To those who have attempted to explore the range of its influence on the nervous system, or to those who look on and note the efforts of others—who see, from time to time, some fresh discovery extend our knowledge in directions altogether unsuspected, and see region after region of disease opened out before the mental eye as part of this affection—the malady may seem like some " dark continent," not yet half traversed, in which the known may even now be less than the unknown. But the nervous system is only one out of many parts of the human frame in which this terrible disease plays havoc, and it may be that the greater frequency and extent of its influence here is merely apparent, due to the readiness with which these delicate structures are deranged in their structure and function, and of the distinctness with which such derangement is revealed. Strange indeed it is to think that all this train of morbid processes—so long, so varied, often so disastrous—is the consequence of the entrance into the system of a few organisms, scarcely to be seen even with the utmost increase in the power of vision that human ingenuity has yet contrived. And strange is it to think that this malady, rivaled in its total capacity for wrecking human happiness, and health, and life, by no other, and exceeded in apparent dreadfulness only by those whose effects are more sudden, is equally formidable by reason of our limited power over it. Make what deductions you can for the mild or latent forms of the disease, and for our power of repression, the fact remains that we have yet to find the means of arresting it; and, I may add, we have yet to find effective means for its prevention. That philanthropy, the vision of which is so narrowed by misplaced feeling that only one imperfect aspect of the means employed can be perceived, has decreed that the chief method hitherto suggested shall not have even a trial under fair conditions, and, despite the manifold benefit from its partial use, has decreed that the malady shall continue to work its ruin on the innocent and the guilty alike. One means alone remains, old as the malady itself, by which it can be prevented. One method, and one alone, is possible, is sure, and that one is open to all. It is the certain prevention secured by unbroken chastity. Is this potentiality increasing? As we look back through the long centuries, we see the sensual more and more dominant as we recede, and clearly lessening as we return toward the present. But when we look around, we can trace small ground for hope that the disease will thus be materially reduced, unless or until there is some change in men more potent and effective than the slow "live upward, working out the beast" of moral evolution. But that which, perhaps, may not be for the mass may yet be for the individual. And, in ending, I must ask a question that I would fain had left unasked, unthought. Do we do all we can—and our profession gives us power that no other has—do we do all we can to promote that perfect chastity which alone can save from this and from that which is worse? The opinions that, on pseudo-physiological grounds, suggest or permit unchastity are absolutely false. Trace them to their ultimate basis, and they are groundless. They rest only on sensory illusions, one of the many illustrations of a maxim I have often to impress on various sufferers, "There are no liars like our own sensations." Rather, I should say, they rest on misinterpretations of these, always biased, and often deliberate. With all the force that any knowledge I possess, and any authority I have, can give, I assert that no man ever yet was in the slightest degree or way the worse for continence or better for incontinence. From the latter all are worse morally; a clear majority are worse physically; and in no small number the result is, and ever will be, utter physical shipwreck on one of the many rocks, sharp, jagged-edged, which beset the way, or on one of the many banks of festering slime which no care can possibly avoid. Let us, then, with our power for good or evil, beware lest we ever give even a silent sanction to that against which, I am sure, on even the lowest grounds that we can take, we should resolutely set our face and raise our voice.

INDEX.

A.

Analysis of fifty cases of syphilitic disease of the cerebral arteries, 71, *et seq.*
Aneurism, intracranial, syphilitic and non-syphilitic, 19, 20.
Arterial disease, syphilitic, seat of, 18–21.
Atheroma, difference between it, and correspondence in seat with it, and syphilitic arterial disease, 18, 19.
Atrophy, chronic muscular, relation of, to syphilis, 50.

B.

Barlow, reference to, 24.
Broadbent, reference to, 10.
Buzzard, reference to, 10.

C.

Cerebral embolism, 70, 77–78.
    hemorrhage, due to syphilitic vascular disease, 21.
    meningitis, chronic, local, of luetic origin, symptomatology and diagnosis of, 67–68.
    thrombosis, results of, 69, 70, and 78.
        premonitory symptoms in, 76–77.
Cerebritis, pathologically specific, rare, 22.
Charcot and Gombault, reference to, 22, 40.
Chastity, unbroken, a certain prevention against syphilis, 126.
Convulsions, significance of absence of, attending onset of hemiplegia due to a cortical lesion, 72.
Cranial neuritis, syphilitic, prognosis of, 105.

D.

Dallinger, reference to, 109.
Diagnosis of syphilitic disease of the nervous system, outline of process, 56–65.
Diphtheria and syphilis, analogy between the effects of, 52.

E.

Evidence, question of, as an aid to the diagnosis of non-pathologically specific luetic lesions, 25.
    of sequence as proof of causation, 25–32
    of treatment as proof of causation, 25, 36.

F.

Functional nervous disorders, the origin of improperly attributed to syphilis, 55–56.

G.

Gummata, common situation of growth of, 18.
  general and special symptomatology and diagnosis of 65.
  special prognosis of, 104.

H.

Heubner, reference to, 68.
Hemiplegia, sudden, not due to embolism or injury, occurring between 25 and 45 years, probably luetic, 77–78.
Hill, Berkley, reference to, 30.
  and Cooper, reference to, 109, 114.
Hutchinson, Jonathan, reference to, 33, 48.

I.

Infiltrating growths, prognosis of, 105.
Iodides and mercury, respective power of, in the treatment of syphilis. 115–117.
Irritation, prognosis of symptoms of, 105–106.

J.

Julliard and Pierret, reference to, 24.

K.

Kahler, reference to, 24.

L.

Latent syphilis, frequency of, 34–35.
Law, Colles's, 35.
Luetic lesions, division of into specific and non-special, 16–17.

"Luetic," the term preferable to "specific," when used as a synonym for "syphilitic," 18.

M.

Meningitis, chronic syphilitic, 37.
  prognosis of all forms of, 104.
Mercury and the iodides, prolonged treatment by, dangerous, 118, 119.
  respective power of, in the treatment of syphilis, 115–117.
Myelitis, disseminated, subacute, 24.
  pathologically specific rare, 22.
  relation of, to syphilis, 39–41.

N.

Narcotic cerebral softening from syphilitic arterial disease, 69, 103.
Neuritis, cranial, pathologically specific, 23.
Nuclear palsy, characteristics of, 87.

O.

Ocular palsies, relation of, to syphilis, 46-48.
Optic neuritis, due to syphilomata, 66–67.
  in the diagnosis of syphilitic lesions, 56.

P.

Pachymeningitis, cerebral, 22.
  spinal, 22.
Paralysis, acute ascending, relation of, to syphilis, 41.
  cerebral, syphilitic, symptoms attending onset of, 73–74.

Paralysis, due to specific cerebral arterial disease, variation in degree of, 73–74.
  general, of the insane, hemiplegia in, 83–84.
  relation of, to syphilis, 49–50, 83–84.
Pathology, ultimate, of syphilis, 12, 51–54.
Percentage of males who have had syphilis, 29–32, 35.
Prognosis of luetic disease of the nervous system, essential principles underlying it, 95–104.

R.

Reason for lack of diagnostic significance of result of treatment in syphilitic cerebral thrombosis, 75.

S.

Sclerosis, focal, of nerve-centres, relation to syphilis, 42.
Sequence, isolated instances of, no proof of causation, 26–27.
"Specific" lesions, character, situations and effects, 18–19.
  wrongly used as euphemism for "syphilitic," 17–18.

Strümpell's hypothesis, 53.
Sylvian artery, that by which arterial disease causes symptoms, 73.
Syphilis an incurable disease, 107–109.
  ultimate pathology of, 12, 51–54.
Sybilitic processes affecting the nervous system, symptomatology and diagnosis of, 65 *et seq.*
Syphilomata, diagnostic value of, effects of treatment on, 67.
System degenerations, relation of, to syphilis, 44–46.

T.

Tabes dorsalis, relation of, to syphilis, 44–46.
Tabetic palsies, nature of, 87.
Therapeutic test, limitations of, availability of, in syphilitic chronic cerebral meningitis, 68.
Tissue formations due to syphilis, pathology, position and character of, 13–16.
Treatment of syphilitic inflammations, better by mercury than by iodides, 117.

www.ingramcontent.com/pod-product-compliance
Lightning Source LLC
Chambersburg PA
CBHW020109170426
43199CB00009B/461